Praise for *The S*

T0090163

"Michael D'Antonio examines the   of [the State Boys'] ordeal in his engrossing book . . . Compelling . . . D'Antonio's book is both engaging and valuable. His State Boys are fascinating people who maintained their humanity and pride against the daily assaults of institutional life. He renders them as vivid individuals, and the warmth of his plainspoken prose makes their stories irresistible."

—*The Washington Post*

"D'Antonio . . . is an exceedingly able storyteller . . . A gripping story."

—*Chicago Tribune*

"It is a vivid, careful, and ultimately momentous piece of journalism, the kind that can set a country back on the right path."

—*The Boston Globe*

"This book ought to be read carefully by all of us who are parents and work with children in schools or hospitals. Here is medical and social history writ large—a melancholy story of vulnerable, needy youngsters become the collective instrument of arrogance and ignorance masquerading as applied science, and the story, also, of children become wondrously brave and determined and knowing critics of state-sanctioned cruelty. Here, too, is lucid, compelling, thoughtful writing—a narrative we readers will ponder long and hard."

—Robert Coles

"*The State Boys Rebellion* is an important and moving story, secret too long. This is why men and women become writers, to tell stories like this, to remind us of who we are at our worst, and at our best."

—Richard Reeves, author of *President Nixon*

"We usually think of eugenics as a social policy that arose in Nazi Germany. But the United States built eugenic institutions where the unwanted were put away long before Hitler rose to power, and, as Michael D'Antonio reveals here, we kept these institutions operating long after World War II ended. This is a brilliant, masterfully reported story, and by recounting the lives of the boys incarcerated at Fernald State School in Massachusetts, he has written a deeply moving book of lasting historical importance."

—Robert Whitaker

"D'Antonio's narrative strikes an admirable balance between the larger social context and scientific theories . . . and the children's lived experience."

—*The New York Times Book Review*

"Luminous writing . . . D'Antonio's analysis of the dark, unintended consequences makes this not only a fascinating read, but a necessary one for anyone interested in how terrible harm can sometimes be born of a sincere desire to do good."

—*The Christian Science Monitor*

"D'Antonio deftly combines detailed archival research and extensive personal interviews to paint a richly nuanced picture of a horrifying and shamefully underexposed part of our country's recent history."

—*Publishers Weekly* (starred review)

"This chilling account leaves you with a sense of wonder . . . Heartbreaking."

—*The Hartford Courant*

"Without exploiting his subjects, D'Antonio pulls back the covers on a world most of us didn't know existed."

—*Star Tribune*

"Insightful, often heart-rending . . ."

—*Boston Herald*

" *The State Boys Rebellion* is the kind of book that makes you give those conspiracy theorists a second listen."

—*Chattanooga Times Free Press*

THE
STATE BOYS
REBELLION

Michael D'Antonio

SIMON & SCHUSTER PAPERBACKS
New York London Toronto Sydney

Simon & Schuster Paperbacks
Rockefeller Center
1230 Avenue of the Americas
New York, NY 10020

First Simon & Schuster paperback edition 2005

SIMON & SCHUSTER PAPERBACKS and colophon are registered trademarks
of Simon & Schuster, Inc.
For information about special discounts for bulk purchases,
please contact Simon & Schuster Special Sales at
1-800-456-6798 or business@simonandschuster.com

Designed by Kris Tobiassen

Manufactured in the United States of America

10   9   8   7   6   5   4   3   2

The Library of Congress has cataloged the hardcover edition as follows:
D'Antonio, Michael.
The state boys rebellion / Michael D'Antonio.
p.   cm.
Includes bibliographical references.
1. Walter E. Fernald State School. 2. Boys—Institutional care
—Massachusetts—Waltham—History. 3. Children with mental
disabilities—Institutional care—Massachusetts—Waltham—History.
4. Inmates of institutions—Abuse of—Massachusetts—Waltham—
History. 5. Child abuse—Massachusetts—Waltham—History. I. Title
HV995.W262W354 2004
362.196'8'00834097444—dc22   2003065741
ISBN 0-7432-4512-1
ISBN 0-7432-4513-X (Pbk)

Photo Credits
1: Courtesy of http://galton.org edited by Gavan Tredoux; 2: *The Black Stork* (Oxford
University Press, 1996); 3: Courtesy Department of Specal Collections and University
Archives, Stanford University Libraries; 4–6: Courtesy of the American Philosophical Society;
7: Courtesy of the State Historical Society of Wisconsin; 8–25 and 27: Courtesy of the author;
26: © Corbis/Ed Quinn; 28: © B. D. Colen/ADIOL.

ACKNOWLEDGMENTS

More than most books, this one has been blessed with the generosity, kindness, and wisdom of those who lived the drama and wanted their story told. Among the former state wards who offered me time and support, I must acknowledge the extra efforts of Fred Boyce, Joseph Almeida, Albert and Robert Gagne, Charles Hatch, Charles Dyer, Robert Williams, Doris Perugini, and Lawrence Nutt. Retired Fernald School staff members also contributed hours of interview time. I was aided especially by Kenneth Bilodeau, Rose Terry, Raymond Pichey, and Lawrence Gomes.

Along with those who once lived at the Walter E. Fernald State School, I benefited from the aid of those who came to know and love them in later years. Doris Gagne and Karen Gagne were especially helpful to me, as was Abra Glenn-Allen Figueroa. Science Club attorneys Wallace Cummins, Mike Mattchen, and Jeff Petrucelly were valuable sources of information and anecdotes. I also received assistance from the staff of the Massachusetts Archives and from the specialists at Harvard University's Countway Library. Sandra Marlow, researcher extraordinaire, freely shared from her trove of documents, books, and audiotapes.

Whatever grace may lie in the pages of this book is due largely to the efforts of my editor, Geoff Kloske, and a team of readers who offered general suggestions and specific criticisms. Ralph Adler was brave enough to risk bruising my ego while demanding better prose. I can say the same for B. D. Colen and Brian Lipson, who pushed me to clarify the history that governed the State Boys' lives. My wife, Toni, and my daughters Elizabeth and Amy, read early versions of the manuscript and made suggestions that made the text more vivid. Fred Wiemer, my copyeditor, smoothed many rough edges.

Finally, special thanks are given here to my literary agent David McCormick, who recognized the merit of the State Boys' story and gave the project the attention and nurturing it needed from its inception. His suggestions for tone, style, and substance are manifest throughout this book.

For Fred Boyce, and all the State Boys.

AUTHOR'S NOTE

The incidents and events depicted in this book have been corroborated by multiple sources, including interviews with participants, contemporaneous notes made by attendants, the records of individual state wards, and reports written by officials of the Commonwealth of Massachusetts. In three instances, the names of persons who appear in this story have been changed to protect their privacy. Each of them is identified in the text by an asterisk.

FOREWORD

After a few months at the Walter E. Fernald State School, seven-year-old Freddie Boyce, skinny with dark eyes and brown hair, could see trouble coming from a distance. He could gauge the mood in the big open ward, and he knew when an attendant was about to lose control. Their job was impossible. No one could handle thirty-six boys, half of whom were retarded, without help. This was why, inevitably, bad things happened.

Some of the boys, the ones who were really retarded, couldn't save themselves even when they were warned. This happened one morning in the fall of 1949, in the minutes after an attendant named Lois Derosier had turned on the lights and walked up and down the rows of iron beds pulling on blankets and barking at everyone to get up. Following the routine, Freddie crawled out of bed and quietly stood in his state-issued flannel nightshirt, waiting to be called to use the bathroom. Everyone else did the same, except Howie, who for some reason felt like wiggling around and chanting, "Whee, whee, whee," over and over.

Howie was mildly retarded—a Dope, in the words of the brighter boys—but he liked Freddie and would listen to him—usually. This time, Freddie began with a quiet "Shhh!" When that didn't work, he tried bribery. "Howie," he whispered under his breath. "If you're quiet, you can sit with me at breakfast."

"Whee, whee, whee," answered Howie.

Derosier, a taut-looking woman in a white nurse's uniform, shouted from across the room. They would all be punished if the noise didn't stop. "Cut it out!" hissed Freddie. "You'll get us all in trouble." Howie just grinned, and continued, "Whee, whee, whee."

"All right!" shouted Derosier. "Red cherries for all of you."

Red cherries were the welts made when the attendants spanked the boys with wooden coat hangers, the same hangers they made in the wood shop. This time, as always, it was done with a certain ritual. The boys lined up at the foot of their beds, pulled up their nightshirts and then pushed down their underpants. Derosier began to work her way down the line. Some of the boys cried when the hanger hit their buttocks. Others took it in silence. Like men, they thought.

It may have been nervousness, or it may have been because he had been prevented from getting to the bathroom after sleeping all night. Whatever the reason, as he stood there with his backside exposed, waiting for Nurse Derosier to get to him, Howie peed. The yellow stream ran down his leg, onto his underpants, and then puddled on the floor. He started to cry even before she saw it. When she did notice it, she became very calm.

"If that's the way you want it today boys, okay."

Derosier left the room. The boys struggled to pull up their underwear. When the nurse returned, she held a big metal bowl. She grabbed Howie and pushed him toward the bathroom.

"You all need to pee?" she yelled, turning to the others. "You all need to pee?"

She ordered Freddie and three others to come with her, too. When they were all in the bathroom standing near the toilets, she held the bowl low and told them to urinate into it.

By this point, Freddie was trembling with fear, anger, and a sense of foreboding. He had never been forced into this position before, where he would be helping an attendant do something awful. He strained to produce a stream. So did the others. When they finished, and the bowl was half-filled, Derosier told them to step back. She then threw the warm contents of the bowl into Howie's face.

Howie's shriek echoed off the bathroom tiles and into the ward. In the bathroom, the boys stood, shocked, until Derosier shouted her four recruits back into the ward. She turned the shower on and told Howie to take his clothes off. When he was rinsed, he and the others went to their cubbies to get dressed. Half an hour later, she cheerily announced "Breakfast!" and led the entire ward downstairs.[1]

*　　*　　*

Lois Derosier's assault on a child in the dormitory of a school for the retarded in suburban Boston remained vivid in Frederick Boyce's memory for more than fifty years, long after he had lost track of Howie and forgiven Derosier in his heart. He would recall it not for shock value, but to show that ideas have consequences. After all, Howie, the other boys in the ward, and Nurse Derosier would never have been together at Fernald if experts and reformers hadn't decided in the early 1900s to perform a great service to humanity.

The Fernald School and more than one hundred other state-run institutions that confined hundreds of thousands of American children—many of whom were utterly normal—were at the center of a long-running but largely forgotten national effort to engineer a better human race. For more than fifty years, physicians and bureaucrats applied the principles of animal husbandry—attempting to weed out bad stock—to troublesome boys and girls. All were branded feeble-minded, or mentally defective, but a great many—including Freddie—would, by modern standards, be considered quite normal.

The American eugenics movement campaigned to build the institutions and gathered up the children deemed unsuitable for reproduction and perfect for commitment. Once a hugely popular and patriotic cause, this effort gained international renown, and contributed directly to the genetic ideology behind the Nazi Holocaust. At home, eugenics' targets became, arguably, the most violated and least acknowledged victims of government abuse in American history. As mere children, they were railroaded into institutions by officials who misused IQ tests. Once locked away, they endured isolation, overcrowding, forced labor, and physical abuse including lobotomy, electroshock, and surgical sterilization. Some of these practices continued as late as 1974 when hundreds of seemingly normal children and adults were discovered at state training centers for the retarded in Florida. Many had been held for more than twenty years.[2]

Though tens of thousands were ultimately released, survivors of America's state schools entered adulthood grossly undereducated or illiterate. They struggled to support themselves and to live independent lives. Many never spoke of their past, because they feared being stigmatized. Some fabricated elaborate stories to hide the truth about their lives from their spouses and children. Those who did talk were rarely believed.

Oddly enough, it was the revelation of a Cold War scientific experiment that finally brought credibility to the small number of men and women who had always been willing to speak about their lives at one state school. The records that were opened as a result of the experiment scandal, when added to their personal stories, provide an accurate and dramatic picture of a place and a time that would have otherwise been forgotten.

But in order to understand what happened to them, one must first grasp the extreme ideology, once presented as scientific fact, that persuaded great numbers of Americans that certain substandard children must be identified, hunted down, and locked away. This distortion of science would live long after its creators, continuing to justify the imprisonment, abuse, and mutilation of the innocent. The most troubling of all the ironies in their story is that it all began with a grand desire to do good.

The state school that became Frederick Boyce's home in April 1949 was America's first institution for the so-called feebleminded. It was founded in Boston in 1848 by Samuel Gridley Howe, a doctor, educator, and husband of suffragist Julia Ward Howe, who was the author of "The Battle Hymn of the Republic." Samuel Howe was an idealist. He believed that with proper attention, children once deemed untrainable could be educated to lead productive, independent lives. His school, which used methods he had observed in Europe, was a success. For a time Howe enjoyed a national following, with parents sending him their children from as far away as Texas.

Samuel Howe died in 1876, but his school continued to thrive. In 1887, the Massachusetts School for Idiotic and Feebleminded Youth moved to a hilly, ninety-five-acre tract in the suburb of Waltham. There a village of red brick buildings rose amid orchards and sprawling vegetable gardens. Shop courses that trained children for factory work were added to the curriculum, and residents did most of the chores that kept the place going. What began as a school became a self-contained community, but the objective was the same: after acquiring basic academic and job skills, residents were expected to graduate to a life beyond the institution.[3]

While the Howe approach to training the feebleminded continued to be effective, expert opinion and public attitudes about them had already begun to change. Darwin's theories of natural selection, and a revived interest in Gregor Mendel's work with genes, led many in the emerging social sciences to argue that human traits, including intelligence, character, and morality, were biologically rooted. From this perspective, the students in Waltham could never be trained adequately for life on their own. Worse, if they were allowed to leave state custody, they would produce many mentally deficient offspring who would become a burden on society.

Inevitably, a rigid embrace of the Darwin-Mendel view would lead to fear that the feebleminded would ruin future generations. One of the earliest alarms about this kind of genetic future was sounded in *The Jukes: A Study in Crime, Pauperism, Disease, and Heredity.* Authored by an amateur criminologist named R. L. Dugdale, this influential book blamed most social ills on a wildly breeding underclass. A slim volume loaded with serious-looking charts and graphs, *The Jukes* excited Victorian-era readers with portraits of harlots, bastards, and vagrants. "Fornication, either consanguineous or not, is the backbone of their habits, flanked on one side by pauperism, on the other by crime," wrote Dugdale. Some of these people could be saved by a change in environment, he concluded, but those affected by "idiocy," which he believed was the result of heredity, were beyond help.

The proposed solution to the problems posed by Dugdale was called eugenics, a movement that sought to apply the principles of selective breeding to human beings. (Based on Greek words meaning "good in birth," the term was coined in 1883 by Darwin's cousin, Sir Francis Galton.) Eugenics declared as fact the superiority of upper- and middle-class white Protestant Anglo-Saxons. It also provided seemingly scientific reasons for Americans to fear the poor and the illiterate as well as immigrants from southern and eastern Europe, Asia, and the Far East. Stripped of scientific jargon and social niceties, the eugenic analysis held that these newcomers, joined with homegrown defectives like the Jukes, would render the nation stupid, lazy, and wantonly immoral.

The eugenics message arrived at a time when civic leaders, academics, and others were troubled by the rapid growth of tenement slums in America's cities. Infectious disease, violent crime, and truancy were problems in these communities. Child labor, sweatshop factories, and prostitution were cited as evidence of a true crisis and harbingers of chaos.

The nation responded with a spasm of reforms that were eventually collected under the political banner of Progressivism. (Devoted to both civic and personal improvement, this was not the same as the far-left progressivism that gained strength after 1929.) In many instances, Progressive social workers and other do-gooders achieved genuine gains for the needy. In other areas, such as immigration, prejudice overwhelmed reason as strict limits were placed on Asian immigration. But all of these campaigns had one thing in common: they depended on experts from the new social sciences, who lent the aura of science to the various causes.[4]

A central element of Progressive ideology, eugenics followed two strategies. The first—positive eugenics—saw the creation of organizations such as the Race Betterment Foundation, which urged white, upper-middle-class women, whose birth rate had declined by half in sixty years, to have large families and devote themselves to their care. Physicians, teachers, and other advocates advanced this idea with public talks, pamphlets, and exhibits at state fairs. One medical journal offered $1,000 to the couple with the "perfect eugenic marriage." Theodore Roosevelt, already concerned about the supposed feminization of the American male, declared that America needed a race of "good breeders as well as of good fighters."[5]

While certain Americans were encouraged to birth a promising future, the other face of eugenics—negative eugenics—fought to keep the genetically unworthy from having babies. Inspiration for this battle was offered by a histrionic book called *The Passing of the Great Race*, which presented Nordic people as a superrace under terrible threat. In a tour de force of pseudoscience, author Madison Grant, who was chairman of the prestigious New York Zoological Society, went so far as to claim that the Roman Empire was "Nordic rather than Mediterranean" and that Greeks were genetically inclined to acts of treason.

In America, "maudlin sentimentalism" had made the nation an "asylum for the oppressed," who were "sweeping the nation into a racial abyss," wrote Grant, who counted President Roosevelt as an admirer. To save the nation, he proposed a "rigid system of selection through the elimination of those who are weak or inferior."[6]

Elimination was actually practiced by eugenic doctors who pressured parents to deny lifesaving medical care to newborn infants with birth defects. This kind of mercy killing was the subject of a 1917 feature film, *The Black Stork,* which argued that God considered euthanasia to be a loving act, and that the souls of these babies would go to heaven and leap joyfully into the arms of Jesus. A few eugenic physicians promoted this practice in interviews, and newspapers published hundreds of articles on the issue. Many included pictures of their patients and their mothers.[7]

The *Black Stork* activities of certain doctors were too controversial to be widely embraced by those who promoted negative eugenics. Instead, they focused on getting laws passed to limit immigration from certain countries, and on suppressing the birthrate among the poor, the immoral, and the unintelligent. Propaganda was a big part of this project, including street-corner rallies where illiterate men were recruited to carry placards with slogans such as, "I cannot read this sign. By what right have I children?" Eventually, politicians responded to the eugenics lobby with policies that turned state schools into asylums where the genetically inferior could be isolated forever.[8]

The question of just who would be confined was taken up by one of Howe's successors in Massachusetts, Walter E. Fernald. Dr. Fernald thought that Howe had been wrong about the potential of the feebleminded and about the school's program. He was certain that the vast majority of the students at Waltham would never be self-sufficient. He was also concerned that the most dangerous among the feebleminded were children who were not obviously deficient and thus might escape official notice. As adults these "brighter feebleminded" would produce "degenerate children" who would perpetuate the social problems personified by the Jukes. "As a matter of mere economy," he wrote, "it is now believed that it is better and cheaper for the community to assume the permanent care of this class before they have carried out a long career of expensive crime."[9]

Fernald and most others in his profession believed that if states were allowed to identify and then remove the almost-normal from the human breeding stock, the whole nation would benefit. This idea would be put into action following two critical events: creation of the IQ test, and the invention of the moron.

A self-made scientist named Alfred Binet produced the technological breakthrough that made eugenics a practical method for human engineering. A lawyer by training, Binet was intensely intelligent, a Parisian who wore pince-nez glasses and sported a pointy waxed mustache. He learned psychology as an acolyte of several French analysts, and rose to become director of the Sorbonne's Laboratory of Experimental Psychology. In 1904, the French Ministry of Education commissioned him to develop a test to identify children who were slow learners. Using elementary schoolchildren as experiment subjects, he eventually settled on a test that presented a variety of tasks including simple math, vocabulary, and puzzles. Those who scored below average were gathered by Binet into two categories: "idiot," for those with a mental age of two and younger, and "imbecile," for those with a mental age of three to seven.[10]

In America, Stanford University professor Lewis Terman translated and refined Binet's original exam to create what came to be called the Stanford-Binet Test. Terman stressed a new interpretation of the scores children produced—the Intelligence Quotient—which was derived by comparing mental age with chronological age. Superior subjects yielded IQs above the mean, which was a score of 100. Inferior children scored below it. For those, Terman held onto Binet's categories, but ordered them by IQ rather than years. Idiots scored below 30 and imbeciles between 30 and 50. He had no name for those scoring between 50 and 100. However, he recognized the importance of identifying such children, so they could be prevented from degrading future generations.[11]

Soon after he developed a means for reducing intelligence to a number, Terman found a way to use it. He compiled a number of IQ studies and reported that when scores were sorted by ethnic group and nationality, nonimmigrant Americans ranked highest, followed by northern Europeans, Italians, Portuguese, and Spanish. These findings supported Terman's belief that intelligence is inherited and therefore as fixed as eye color. He also used them to argue that "educational reform may abandon, once and for all, the effort to bring all children up to grade."[12]

While Terman fine-tuned his test to identify "high-grade defectives," Henry Goddard investigated the threat they posed to human progress. A researcher at the New Jersey Training School, Goddard offered his findings in a best-selling book called *The Kallikak Family: A Study in the Heredity of Feeble-mindedness*, which told the tale of six generations of despair in a family living in the New Jersey Pine Barrens. (Goddard claimed to be able to diagnose the dead through interviews with descendants and by looking at their pictures. The flaws in his methods and the fact that he had altered some of the photos published in his book would not be revealed for decades.)[13]

Though they were too intelligent to fall into one of Binet's two groups, Goddard insisted that the troubled Kallikaks were nevertheless feebleminded, and descended from a single tavern girl who had polluted the family bloodline. No amount of education could save future Kallikaks from feeblemindedness any more than it could "change a red-haired stock into a black-haired stock," wrote Goddard.[14]

With the evil of the nearly normal defined, Goddard sought to give them a name. For this he fashioned the Greek word for "foolish"—*moronia*—into a medical term—moron. (A later name for this group—the Almosts—never caught on.) Morons were more capable than Binet's imbeciles, said Goddard. Some scored as high as 70 on IQ tests. And because their defect was slight, they were likely to escape authorities and commence breeding.

Terman feared that tens of thousands of such high-grade defectives lurked in the countryside, just waiting to deceive and then destroy normal families. Eventually, he inferred, the nation would descend into intellectual and moral mediocrity. The solution Goddard proposed would depend on the dramatic expansion of institutions like his own.

They should become "colonies" for the feebleminded, he said, separating them from society as if they were lepers.[15]

The creation of the moron as a public danger, and an IQ test that allowed them to make certain diagnoses, allowed Terman, Goddard, and their colleagues to offer themselves as America's defenders. State officials began to encourage social service agencies, courts, and police to send them suspected morons for testing. Those who scored below normal would be admitted to state schools. Once locked in an institution, the "high grades" would receive efficient custodial care and training. They would, as a matter of public policy, be prevented from reproducing. As a side benefit, their former schoolteachers would be relieved of their most difficult students.

In language typical of Progressive idealists, Goddard framed the pursuit of the moron in terms that were intended to both alarm and mobilize: ". . . we need to hunt them out in every possible place and take care of them, and see to it that they do not propagate and make the problem worse, and that those who are alive today do not entail loss of life and property and moral contagion in the community by the things they do because they are weak-minded."[16]

To make sure every last moron was captured, many states, including Massachusetts, would establish traveling "clinics" to administer IQ tests at public schools. Often these clinics identified as feebleminded, children who had been deemed by teachers and parents to be normal. Many of these boys and girls would be separated from their families by officials who pressured parents to accept that an institution offered their child the best possible future. Parents who did not volunteer their children for admission often lost custody of a son or daughter in court.[17]

With thousands of children being referred for incarceration, state schools around the country embarked on building programs to accommodate them. Walter E. Fernald expanded the school at Waltham from 400 residents to 1,300. To handle males who grew too old to be among children, he acquired 1,600 acres in rural Templeton, Massachusetts, to serve as their permanent home. And around the country, other superintendents also opened farm satellites, and some entire new campuses.

* * *

The rapid growth of institutions led the editors of the bulletin published at the New Jersey Training School to declare: "When similar colonies exist in every state of the union, and the defectives by the thousands, both men and women, are gathered into them, we shall be beginning to satisfy the greatest of all present social needs, the complete care and control of the defective."[18]

The managers of these institutions would continue to emphasize the need for "sexual quarantine" for decades to come. As late as 1946, the Massachusetts commissioner of mental health would explain that "generally speaking, feebleminded women are potential prostitutes and feebleminded men are potential sex offenders. . . ."[19]

The trouble with the institution solution was that no government agency had the money to build all the housing that would be needed to place every American moron in a secure colony. These limits were evident by the early 1920s, when some superintendents began sterilizing young women so they could be released to the outside world without the ability to reproduce. This solution to the overcrowding of state schools was employed first in California, but by the mid-1920s it reached the East Coast, where a landmark legal case would guarantee it became common in every region of the country.[20]

Dark and slender, with high cheekbones and wavy brown hair, seventeen-year-old Carrie Buck had been raised in foster care, where she was raped, became pregnant, and gave birth, all before the age of seventeen. Photographs of her show a girl with a wary, worn-out appearance. In 1924, after her own daughter was born, Carrie was committed to the Virginia Colony for the Epileptic and Feebleminded near Lynchburg. Eager to begin using sterilization against the breeding of more "degenerates," colony officials went to court seeking permission to surgically sever Carrie's fallopian tubes so that she could be released to live in "liberty under supervision." A new Virginia law had authorized such surgeries, but, before proceeding, the state would use Carrie's case to test the statute's constitutionality.

Lawyers who were old friends represented both the state and Carrie Buck before the circuit court in Amherst, Virginia. An expert witness, who had never even seen her, testified that Carrie should be sterilized because she was part of "the shiftless, ignorant and worthless class of anti-social whites of the South." In fact, Carrie was descended from wealthy landowners and had done well in her limited time in school. She was not even illegitimate, as the state claimed, but rather the daughter of a married couple. None of this background was presented in court, and the judge ruled that the operation could be done.

Appeals to state and federal courts took two years. Finally, in an opinion that recalled the infanticide practiced by the ancient Greeks, Justice Oliver Wendell Holmes ruled for the state, writing: "It is better for all the world if, instead of waiting to execute degenerate offspring for crime, or to let them starve for their imbecility, society can prevent those who are manifestly unfit from continuing their kind. . . . Three generations of imbeciles are enough."

On the morning of October 19, 1927, Carrie was brought to the two-story brick building that housed the infirmary at the Colony. Told only that she needed surgery for health reasons, she was anesthetized and was then sterilized during an hour-long procedure. Years would pass before she learned the truth. In that time she would be released, marry, and retreat into a life of poverty and isolation.[21]

Emboldened by Holmes's ruling, officials in Virginia proceeded to sterilize more than 8,300 people. Thirty states adopted the practice and would together perform nearly 66,000 castrations and tubal ligations on the feebleminded and insane in America. California led the way, with more than 20,000.[22]

Some Americans never accepted the science or social aims of eugenics. By the late 1920s, critics arose to attack the discipline's main tool, the IQ test. Walter Lippmann published several articles in the *New Republic* that exposed the biases in the test and the prejudices of Terman and his supporters. Lippmann wrote that the claim that IQ tests could measure inherited intelligence had "no more scientific foundation than a hundred other fads, vitamins, and glands and amateur psychoanalysis and correspondence courses in will power." One important piece of evidence to support Lippmann's argument was found in the results of IQ tests completed by men mustered into the Army during World War I.

Fully half scored below normal, but instead of seeing this result as a sign that the test was flawed, its proponents used it to inflame the fear that the country was threatened by feeblemindedness.

In the mid-1930s, science began to catch up to Lippmann's essays. Some biologists began to challenge the basic science of eugenics with experiments that showed that most so-called defects were not readily passed from generation to generation. At the same time, new analyses of supposed population differences showed the flaws in eugenics notions about superior and inferior races.[23]

As the biologists picked apart the basic science of human breeding programs, social scientists began presenting experiments that highlighted the role of environment in human development. One of the first studies that challenged the concept of the immutable IQ was published by psychologists Harold Skeels and Harold Dye.

Skeels and Dye had been captivated by the story of two orphan girls, aged thirteen months and sixteen months. Initially diagnosed as familial imbeciles with IQs of 35 and 46, they were placed in a ward of brighter, older women at an institution in Iowa. Surrounded by doting teenagers and adults, the little girls quickly learned to walk and talk, and within a year were found to have normal IQs.

To test their belief that generous amounts of stimulation and care at a young age had made the difference, Skeels and Dye found thirteen orphans under age three, with median IQs of 65, and placed them in wards where they would receive constant attention from more capable older residents. The caregivers became so committed to their roles that they eagerly taught the children to walk, talk, and play games. They even held a Fourth of July baby show, parading the children in carriages. At the end of the yearlong experiment, each child showed an increased IQ, with gains ranging as high as 58 points.

While they watched the group of thirteen improve, Skeels and Dye studied a second set of children with much higher initial IQs— a median of 90—who had been placed in a crowded state orphanage. They received minimal custodial care from nurses. The median IQ of this group fell 26 points during the test period.

Comparing the two groups, the psychologists concluded that "a close bond of love and affection between a given child and one or two adults" could literally raise a child out of feeblemindedness. What they learned about how children declined when neglected—either by incompetent parents or orphanage staff—was equally dramatic: ". . . children of normal intelligence may become mentally retarded" in a "non-stimulating environment," they wrote.[24]

Studies done by the likes of Skeels and Dye affected the views of Walter E. Fernald and Lewis Terman. Eventually, both men would change their minds and argue against locking up ever-greater numbers of morons. But it was too late. The institutions they had helped to create had developed a momentum beyond their influence. State schools continued to grow in both size and number. They were supported in this mission by local, state, and national governments. Under President Hoover, the White House endorsed eugenics as a response to the problem of "defective" children.[25]

Outside government, a network of clubs and associations, nearly all of them filled with intelligent, educated, well-intentioned people, promoted the cause. One of the most important of these was the Eugenics Records Office in Cold Spring Harbor, New York. Funded by prominent New York families that included the Harrimans, Carnegies, and Rockefellers, the ERO promoted research on racial differences, compiled extensive genealogical information on thousands of families, and lobbied on behalf of state sterilization programs and limits of immigration.[26]

American eugenicists also took a leading role in the development of their movement around the world. Germany embraced this science most enthusiastically, with formal organizations for "racial hygiene" dating back to 1904. But from its beginning, the German movement took its cues from America. In 1923, German eugenicists wrote admiringly of the Virginia-style sterilization laws. In 1930, *The Kallikak Family* was published in translation. After the Nazis came to power in 1933, German officials regularly hosted American eugenicists for exchanges of ideas and even honored Harry H. Laughlin of the ERO for his contribution to "the science of racial cleansing." After World War II began, American eugenics proponent Lothrop Stoddard, whose writings appeared in Nazi textbooks, had an audience with Hitler himself. (The Führer had earlier written to America's

Madison Grant to tell him that his book *The Passing of the Great Race* was his "Bible.")

The admiration was mutual. Stoddard came home to praise the Nazis' programs, especially a Hereditary Health Supreme Court, where he saw four people, including a Jew, a manic-depressive, a deaf mute, and a feebleminded girl sentenced to sterilization. Even before Stoddard, the *New England Journal of Medicine* had taken notice of the Nazi programs. "Germany is perhaps the most progressive nation in restricting fecundity among the unfit," wrote the journal's editors.[27]

As investigators later documented, the Holocaust began with the murder of thousands of disabled men and women, including so-called morons. They were killed in rooms designed to look like gang showers but equipped to accept the exhaust from a car engine. These chambers were prototypes for the more efficient killing rooms that would eventually be used to exterminate millions.[28]

After the war, when the horror of Nazi racial policies became known, eugenicists in the United States sought to downplay their connection to the Reich's policies. A 1946 article on the history of the American Eugenics Society made no mention of the prewar relationship between the movements in the two countries. According to German sociologist Stefan Kühl, the Americans promoted the idea that only a fringe supported the Nazis. While this was not true, this view became accepted, and the U.S. contribution to Nazi ideology was lost in the torrent of history.[29]

Although the German-American relationship could be covered up, the Nazi taint made the word "eugenics" a liability. In 1945, the *American Journal of Mental Deficiency* dropped the category called "Eugenics" from its table of contents. The word quickly left the lexicon of medicine, education, and social welfare, and eventually even well-educated Americans would have trouble defining it. But the programs created in the name of eugenics continued uninterrupted. States aggressively identified so-called morons, institutions grew larger and more numerous, and the practice of sterilization continued. In 1949, one booster estimated that surgery had prevented the births of 19,000 mentally deficient children and saved the American taxpayer $117 million.[30]

With so many supposedly subnormal people being sterilized, the eugenicists should have reduced the number of low-IQ children being born, and the population of institutions should have stabilized. Instead, the number of children who were locked away continued to grow, and buildings on state school campuses became overcrowded. This happened because the criteria for the diagnosis "moron" had become looser with time. Where once state schools refused admission to anyone who scored 70 or higher on an IQ test, the 1940s found many children with such scores being labeled and committed as "borderline" cases.[31]

Although some families resisted giving their children up to the doctors, educators, and others who ran state schools, few protests were made on behalf of orphans or abused or delinquent children. Many people believed that experts could do a better job of preparing these children for life in a complex world. Some even wondered if an upbringing in a carefully controlled institutional environment might not be superior to life in the typical American family.[32]

Serious dissent arose from just a few critics, who began to suggest that the rising number of admissions of borderline children was a matter of expediency, a convenient way to remove them from public view. In 1947, George Stevenson, who headed the National Committee for Mental Hygiene, wrote, "Many a social service agency today is relieved to get a diagnosis of mental deficiency on a troublesome case, so that it can unload it on an institution."

Others challenged the basic concept of morons as a group in need of special attention. In 1948, Dr. Ruby Jo Reeves Kennedy of Connecticut College for Women reported that the "typical moron" man in society earned higher pay than the average man of normal intelligence. He was married, had one child, and lived in a home with both a radio and a telephone. The implication was clear: despite being labeled deficient, he was like everyone else.

The doubts raised by Kennedy and Stevenson were not widely shared. By 1949, the year Frederick Boyce was admitted, the population at the Fernald School was nearly 1,900. Across the nation, eighty-four institutions housed a total of 150,000 children, and twenty-six more state schools were under construction. Fernald was about to be expanded, even though officials in Massachusetts acknowledged at the time that about 8 percent of the children in its state schools were either almost normal, or not at all retarded. This figure suggests that nationwide, at any given time, more than 12,000 American boys and girls of relatively normal intelligence were locked away.

Whether they might ever be freed was an open question. Parole was occasionally granted to the most competent adult residents of institutions. Those who were most likely to receive this kind of release had families to take them in, but even in such cases, discharge was a rarity. In 1940, the superintendent of Fernald had explained to the school's trustees that it was far more likely a child would be "here for an indefinite stay."

Superintendent Ransom Greene offered many reasons to explain why so few people left Fernald for the outside world. Among them was the institution's need for the unpaid labor of brighter inmates, who worked as custodians, caretakers, gardeners, cooks, etc.

"The upper level of mental defect helps in the care of those less able to care for themselves, or we would have a very much larger employee roster," noted Greene. To keep the place running smoothly, the trustees decreed that Greene adjust admissions to make sure that at least 38 percent of the Fernald population was composed of higher-functioning children and adults. (People who today would be regarded as normal.) Officially identified as morons, these residents of Fernald were direct historical descendants of the Kallikaks. Pariahs in the outside world, prized for their labor on the inside, they found that the institution became both their present and their future.[33]

ONE

The snow stopped at about three o'clock in the afternoon. Four boys, aged six and seven, pulled on coats, boots, hats, and woolen mittens and scuffled out to the barn, where a wooden toboggan was propped against a wall. They grabbed the sled and headed for the coasting hill, which was a quarter-mile away, through a leaf-bare apple orchard and a stand of pine trees. The cold froze their breath, and the snow muffled the sound of their voices as they argued over who would get the first ride.

When they got to the hill, Freddie, Gordon, Wally, and Foxy stopped for a moment and looked down the slope toward Hoyt's dairy farm. (Freddie, the smallest and youngest of the boys, was frightened by the steep incline, but he didn't show his fear to the others.) In the distance, where the slope ran out, the property line was marked by a stream called Silver Brook, and by twisted strands of rusted barbed wire strung on three-foot-high posts.

On the first few runs, the toboggan barely made it through the fluffy powder to the bottom. But as they kept at it, sliding down the same path and trudging up beside it, the boys made a track of packed snow. The toboggan flew faster with each trip down. It also ran a few feet farther, toward the stream.

As dusk fell, the boys' cheeks turned red, and they panted as they climbed the hill. Pea-sized clumps of snow hung from their soggy mittens and collected on their socks. Their toes and their ears were numb, but under wool caps their hair was soaked with sweat. They had orders to be home by dark, so there was time for just one more ride. Gordon and Wally climbed onto the toboggan behind Freddie, who had already tucked himself under the sled's curved wooden front. Foxy gave them a big push.

With the air growing colder by the second, the track was turning to ice. As the toboggan hurtled forward, Wally and Gordon realized that it was not going to stop before they got to Silver Brook. They tumbled out and yelled for Freddie to jump. Freddie didn't respond. When he finally stirred, it was to lean to the side, steering the sled away from the brook and into the barbed-wire fence.

Freddie's face was so numb from the cold that he didn't feel any pain when the rusted barb caught his right cheek. But he did feel a pull, like the tug a fish must feel from a hook and line. When his flesh tore, the wire snapped free. The toboggan skidded to a stop. Freddie rolled out and then sat up. He looked down to see red drops on the snow. The other boys, who had run down the hill, pulled him to his feet and began to walk him back up the slope.

When they reached the barn, the warm blood finally thawed Freddie's face and the nerve cells in his cheek began to scream. He cried and the boys shouted for help. Inside the house, sixty-seven-year-old Marion Bond heard them and went to the door. She brought the boys inside, put a hot, wet cloth to Freddie's cheek, and told him to hold it there, firmly, while she telephoned for a doctor. By the time she returned, Freddie was still in pain, but he was more concerned about what Mrs. Bond was going to say and do. She sat down on a chair and hugged him hard, trapping the wet cloth between his face and her breast. He noticed she was warm and smelled like soap. When she let him go, she continued to press the bandage against his cheek, to stop the bleeding. The room smelled like wet wool and sweaty boys.

Freddie didn't cry when the doctor arrived and tugged the blood-encrusted cloth away from his wound. He stood bravely as his skin was stitched closed, and he even stayed calm for a tetanus shot. Freddie had always had the ability to retreat into his mind, shutting out whatever physical or emotional pain was at hand. This skill allowed him to feel good about the attention he was receiving. He would forever remember the doctor's visit and his run-in with the barbed wire as a positive experience. The warm cloth, Mrs. Bond's embrace, and even the doctor's needle felt like love.[1]

* * *

Frederick Boyce had followed a loveless path to the Bond farm on Hadley Road in rural Merrimac, Massachusetts. He had been born in Boston on January 12, 1941, to a mother who had just turned twenty-one. A short, skinny, dark-haired woman with little education, Mina Boyce had drifted through life, transferring her dependency from one unreliable man to another. When Freddie arrived, she already had one child, a two-year-old named Joseph, who state records noted was "illegitimate." She had a problem with alcohol, and she was a widow. Her husband, a steamfitter from rural Maine, had committed suicide before her second child's birth.

Mina had held on to her sons until August of 1941, when neighbors called police because the children had been left alone in her apartment in Boston. Social workers from the state's Department of Public Welfare came with the police, who broke open the locked door. The officials took custody of the children, placing them with separate foster families who were overseen by the department's Division of Child Guardianship. Frederick, who would never see Joseph again, landed in a foster home in the small town of South Easton, about thirty miles south of Boston. There, under the care of Mrs. Kathleen Brophy, he survived bouts of grippe and scarlet fever. He learned to walk and talk, but speech came slowly. He could say just a handful of words. Small and thin for his age, Freddie had dark skin, brown eyes, and thick curly hair that was turning from light to dark brown.

The house in South Easton was filled with kids. Some, like Freddie, were wards of the state. Others were Mrs. Brophy's biological children. She kept control with harsh discipline and threats. Like most children, Freddie learned the power of saying no at age two. He tried it out a few times with Mrs. Brophy, and she warned him against it. He became much more cooperative when he saw her snatch up a very young child who had wet his pants and dunk him, feet-first, in a flushing toilet. The punishment was designed to teach the children about the importance of potty training and obedience. Freddie got the message.[2]

As an adult, Fred Boyce would recall that in the Brophy home the state wards ate separately from the family. Their diet was heavy on spaghetti, potatoes, and cereal, and light on meat, milk, and eggs. Freddie never got enough, except once. It happened on a night when other boys had sneaked into the kitchen and stolen food. When they were caught, Freddie was awakened, brought to the kitchen, and required to eat a large meal while the others looked on. Confused and frightened, he gobbled the food quickly and then threw up.[3]

In the two years that he lived with Kathleen Brophy, Freddie grew into an active toddler, but he was especially shy. Though he didn't understand why he received less attention than Mrs. Brophy's biological children, he surely felt the difference. He seemed afraid to talk to adults, and other children had trouble understanding him. He breathed heavily, and at night he snored like an old man. These problems improved when he had his tonsils removed. A few months later, he made his first visit to a barber, where his soft, brown curls were cut off. The next time a social worker visited him, she noted that "he has lost his babyish look."

State records show no reason, but Freddie was moved to another home, this one in the town of Hingham, in October of 1944. Two days after Christmas, when a social worker came to that house, Freddie ran to hide, afraid he was to be uprooted again. He was right. The "state lady," as he called her, packed his clothes into paper sacks and brought him, resisting all the way, out of the house to a car. She put him in the back seat, along with his things, and drove away. In an hour, he was at yet another foster home—his third—in the town of Dedham. A few days after his arrival, he turned four years old.

The new foster mother, Margaret Forrester, worried about Freddie's speech problems. Though he seemed aware of everything that was said, he spoke in a strange, tight-mouthed way that was often unintelligible. Among the few words he said clearly were "cat" and "mama" and "barber." (Evidently, that first haircut had made a big impression on him.) For the first time, a social worker recorded a note that suggested that this skinny dark-haired boy with penetrating eyes was not being well served by the Commonwealth of Massachusetts. She wrote:

> Foster mother: Frederick was in bad shape when she got him.
> He was thin. Child would sit for hours without moving, and
> she discovered that he had some sores on his head.

Freddie stayed nearly three years in this place, making his first true friend—a boy named Robert—and showing his foster mother that he was bright, even advanced for his age. He cleared the table after meals, dusted Mrs. Forrester's living room when she asked him to, and helped to watch after the younger children. He understood complex instructions and was a peacemaker with the other boys when things got rowdy. Still, his garbled speech troubled adults. A social worker reported:

> Foster mother persuaded F. to sing a song. He did this by barely
> opening his mouth.

Based on caseworker recommendations, the state sent Fred to a physician who found him capable of talking, but reluctant to engage in extended conversations, and he predicted the boy would "grow out of it" without any special help. Otherwise, the doctor found that Freddie was a normal boy.

The social workers assigned to his case did not accept the doctor's evaluation of Freddie Boyce. Still troubled by his speech, they were determined to discover the cause of his problem. In July of 1946, a state lady took him to an institution for the retarded in Wrentham for testing. (Located in the town by the same name, the Wrentham State School was about five miles northeast of the Massachusetts border with Rhode Island and a half-hour's drive from Mrs. Forrester's home in Dedham.)

For the intelligence test, he was required to repeat five-digit numbers, to define words such as "timid" and "tame," to name colors and shapes, and to answer questions about stories. One story used for children his age in that time was titled "The Wet Fall." After listening to eight short sentences that told the tale, children were asked to repeat the title. Those who said it right, or said, "A Wet Fall" or "One Wet Fall," or even "A Wet—something about a fall" received credit. Those who said, "The Wet Falls" or "A Fall" lost points.[4]

Freddie strained to please the adults he met at Wrentham and to get every question right. But since he had never attended school, he was puzzled by much of what he was asked to do. He had rarely seen books and had no experience at all with pencils or crayons. Confronted by strangers who used words he didn't understand, and who asked him to perform task after task, he grew increasingly anxious and more dependent on the state lady. Between sessions, he would go into a hallway, climb onto the bench where she sat, and cling to her for security. This behavior would be noted in his test report as "regressed" and "immature."[5]

Two days after the testing, the experts at the Wrentham State School declared Freddie Boyce "feebleminded of the familial type." In a brief report, they noted that his IQ was 65, and his mental age was two years behind his chronological age.[6]

Back in Dedham, Margaret Forrester, who didn't want to care for school-age children, prepared to get rid of Freddie Boyce. She began telling him, "You're going to have a new mama next year." In the spring of 1947, Freddie watched as his pal Robert was taken away. Convinced that he was next, he became sullen and withdrawn. The progress he had made in talking stopped, and he seemed to forget some of his new words. Every time a social worker came to visit, he cried or tried to hide.[7]

> As soon as F. saw visitor his eyes filled with tears and he began
> to whimper. He sees visitor and thinks that he is to be moved.
>   Visitor explained to him that she had merely come to see
> how he was getting along and did not come to move him.

Given his recent diagnosis by the doctors at Wrentham, supervisors at the Division of Child Guardianship concluded that Freddie belonged in a state-run residential school for retarded children. But their application was denied, because the institution was full. With Mrs. Forrester still demanding that Freddie be moved out of her home, he was finally transferred to Mrs. Bond's farm in Merrimac. It was his seventh home in six years.[8]

Tall and thin, with gray hair that she often tied in a bun, sixty-six-year-old Marion Bond could have passed for a woman ten years younger. She wore pretty dresses, and her face was smooth and digni-fied, but she was so strong that when a drought emptied her well she could carry all she needed from a neighbor's house without any help.

Childless, and a widow for fifteen years, Mrs. Bond had maintained her home but let the farm go. Her fourteen acres of fields and orchard were slowly turning wild. The roof of the barn, which was built at the time of the American Revolution, was so leaky that when it rained there was hardly enough dry space for the one milk cow she still kept.

Mrs. Bond survived on the two crops a woman her age could raise without help—September hay and foster boys. Each year when the hardwood trees turned color, a neighbor came up the road from his dairy farm with a horse-pulled mower to cut the hay. The foster boys watched from the barnyard as the whirring blades threw dust, slivers of grass, and insects into the air. After the hay was stacked, they jumped into the sweet-smelling piles.

If Marion Bond saw the haying operation as the last vestige of real productivity on her farm, the boys saw it as an example of how the place pulsed with life. They had all come from crowded houses and tenement apartments where they had slept with hunger almost every night. In Merrimac they picked the apples, cherries, and plums that still grew on the farm and ate as many as they could.[9]

That Mrs. Bond was generous with good food was enough to make her the best of the six foster mothers who had taken custody of Frederick Boyce in his short life. But like the others, she was not affectionate in her day-to-day contact with the boys. She kept them clean, and healthy, but she did not read to them or tuck them into their beds at night.

She was strict about her privacy, and confined the boys to their rooms, the kitchen, and the enclosed porch where they took meals and played. Strict as she was, the boys meant more to her than the $20 per month she earned taking care of each of them. She gave them ice cream every Sunday and celebrated their birthdays. She told them they were the equals of every child in town, and sent them to church and to school so they could see this was true.

In September 1948, Freddie and the others rode a yellow bus to Merrimac Port School, which overlooked the Merrimac River as it flowed through the center of town. In first grade, Freddie struggled with his lessons, but earned one of the warmest memories of his childhood when a teacher noticed he had forgotten his lunch and shared her sandwich with him.

Marion Bond's boys spent six weeks in their classes before the local school board, which served a middle-class community of farmers and factory workers, voted to remove the state wards from their classes and bar them permanently. The reason they gave was that the boys were not regular town residents and they needed expensive, special instruction, which was not available. Superintendent John C. Paige informed Mrs. Bond and the state that the decision was final, and he recommended the boys be sent to a city with a larger school system that might accommodate them. The state responded by sending a social worker to meet with Paige. He resisted her appeal, and according to Freddie's records, no effort was made to put the boys in another district.[10]

All that fall, while neighbor children sat in classrooms, the boys at the Bond farm busied themselves outdoors. Though they must have been aware of the fact that they were missing something that other children got in school, they found plenty to do around the farm. They dug in mounds of earth and found ancient arrowheads, which made their games of cowboys and Indians feel more exciting. They also explored the old barn and climbed trees. Though they were free to play much of the time, the boys at the Bond farm also did small chores. When the potato crop came in at a neighboring farm, some of them helped pick. They filled bushel baskets, earning a few pennies per hour.

On Halloween, Marion Bond let the boys stay outside extra late. One of them thought it might be fun to light the night with torches. Hay was bound to sticks, and stolen matches were struck. The boys hoisted the torches high and ran through the night until they frightened themselves. One of them dropped his torch and set the dry stubble on the field ablaze. From a distance, as the boys ran about in confusion, they must have looked like pagans dancing around a ritual fire.

When she saw the fire, Marion Bond screamed for the boys to come to the house. She made sure each of them was safe before she called the fire department. One engine responded, and the fire was quickly extinguished, leaving behind a circle of blackened dirt about sixty feet across. After the fire company left, Mrs. Bond lectured the boys on matches and fire, but no one was spanked or otherwise punished. They had already upset themselves enough.[11]

When winter came, Freddie and the others spent more time indoors. A wood-fired, cast-iron stove warmed both the kitchen and the enclosed porch where the boys drew pictures with crayons, played checkers, and pieced together puzzles. It was Freddie's first exposure to these kinds of playthings, and he liked everything about them, including the waxy smell of the crayons and the smooth edges of the puzzle pieces.

The run-in with the barbed-wire fence would have been Freddie's most vivid memory from that winter at Mrs. Bond's house if the state had succeeded in finding him a place in an institution. But because their efforts failed, he awoke in the farmhouse on the morning of March 1, 1949, to a strange silence. Normally, he would have heard the clang of pots and pans, and smelled coffee. He would have felt Mrs. Bond's footsteps before she called him to breakfast. On this morning, however, the house was quiet and cold.[12]

One of the boys knocked on Mrs. Bond's bedroom door and discovered that she was not well. Another called the police, and soon a squad car, an ambulance, and then the cars driven by the social workers filled up the barnyard. Clothes were collected in paper bags, and the boys were seated at the table on the porch. They were told that Mrs. Bond was dead, and then one by one they were taken off in separate cars. Each one of them was crying over losing Mrs. Bond and the home she had made for them.

Seated in the back seat of a state car, Freddie rode about forty minutes to another small town, Bradford, and a house where he shared a room with another boy for seven weeks. There he felt even lonelier because his attachment to Mrs. Bond had made him feel, for the first time, that he belonged to some place and to some person. Losing her filled him with the sense that he would never again feel at home anywhere. He didn't mention these feelings to anyone and was determined that no one would see him cry. But his days were filled with a hollow ache, and at night he sometimes awoke drenched in tears.

On April 26 another social worker came to pack up Freddie's things. She dressed him in clean clothes and told him to put on his leather shoes. She then drove him to a courthouse that was near the golden-domed State House in downtown Boston. Though summoned, Freddie's mother did not appear. If she had, it's unlikely that her seven-year-old son would have recognized her since she had never once visited him since he became a ward of the state at eight months of age.

The hearing began with a representative of the Division of Child Guardianship recommending that Freddie be committed to a state institution. A psychiatrist certified that Freddie was, indeed, suffering from a mild level of retardation, which he termed "feeblemindedness." He said that a state school, which was at last ready to accept him, would be the best place for him. No one actually spoke for Frederick Boyce. This was the era before children received independent, court-appointed legal guardians. A judge signed an order committing Freddie to an indefinite term at a state school.

Although he had heard every word of what happened in the court-room, Freddie understood little. He didn't know what "feebleminded-ness" was, and he couldn't fathom the meaning of a commitment order. All he really understood was that he was completely powerless and that he had no choice but to trust the strangers who surrounded him, talked about him, but never addressed him directly.

Outside the courthouse, the social worker took Freddie's hand and led him on a short walk to the Park Street subway station. Having never seen the subway, he walked carefully down the steps and stood tentatively on the platform. The screech of the metal wheels on the track was deafening, and if the state lady hadn't tugged on his hand, he would have been too frightened to board the car. At North Station, they transferred to a suburban line. Soon after it left the station, it flew out of a tunnel and into the daylight. It crossed the Charles River and headed into Cambridge and points west.

From his seat, Freddie could see only the utility poles that whizzed by outside as the train rushed to the suburbs. He was mesmerized as they passed in a rhythm, looking black against the sky. It was a little after two o'clock when the train slowed for Waverley Station and the social worker took Freddie's hand. She pulled him to his feet, and tugged him toward the door. When the train stopped moving and the door opened, they stepped down and walked across the platform to a

taxi. The cab—another first for Freddie—pulled away from the curb
and turned east on Route 60 and then north on Trapelo Road, traveling
past comfortable single-family homes crowded onto small lots. In less
than ten minutes the car slowed, and the driver put on the blinker to
turn left. When the way was clear, he turned to head up a long narrow
lane.[13]

The boy in the back seat sat up straight to look out the window. To
the right, he saw an emerald baseball diamond, with dirt base paths
and a worn pitcher's mound. Beyond it rolled acres of farmland, fur-
rowed but not yet planted. On the left, a green lawn ran up a hill
topped by a cluster of imposing brick buildings.

Planted near the roadway, at the entrance to this little world of
green grass, red brick, and brown earth, a white wooden sign with
black lettering he could not decipher announced that Freddie had
arrived at THE WALTER E. FERNALD SCHOOL FOR THE FEEBLEMINDED. As
he passed the sign, the boy turned and looked out the rear window and
saw the cars passing on Trapelo Road. They carried other people, perhaps boys like him, to destinations unknown in a world he would
come to call simply "the outside."

TWO

The taxi climbed the slope toward the Fernald State School Administration Building, three stories of brick with big windows and white columns and a clock tower sprouting from the center of its gray slate roof. Freddie stared at the imposing structure. It reminded him of the church steeples in Merrimac.

At the top of the hill, the car turned left, down a tree-lined lane, toward a cluster of ornate brick buildings arranged around a grassy courtyard. When it reached the yard, the state lady told the driver to stop and wait. She took him by the hand and led him down a concrete walk, toward the building called the Boys Dormitory.

From the outside, the Boys Dormitory seemed part castle and part red brick schoolhouse. The center section of the façade was outlined with a high masonry arch, which sheltered bow windows on each floor. A decorative cornice flared from the top of the building to support wide copper gutters that had turned powdery green. Four large ventilating shafts, also made of brick, rose like turrets from the midsection of the roof.

Freddie and the social worker walked down a half-dozen steps to the large wooden door, which opened onto the center hall of the building. The office for the dorm was on the left. Freddie sat alone on a bench in the hallway while the state lady went inside.

From where he sat, Freddie could see a wide wooden stairway. He could smell food cooking. He heard footsteps as a solitary boy came down to the landing of the stairway, peeked at him, and ran back up. On the train, and again in the car, the state lady had told Freddie that he was going to a big house where lots of other boys lived. Now he was a little curious.[1]

When the state lady emerged, she took Freddie inside to where a nurse in a white dress sat at a typewriter. Her name was Eleanor Jacques and she peered at him over her glasses. In her hands, she held a shirt, a pair of denim overalls, heavy black shoes, and thick black socks.

The nurse agreed to let the social worker stay a bit longer. She handed her the clothes. "Make yourself useful and get him into these," she said. Right there in the office, with the door wide open, the state lady stripped Freddie and then dressed him. He would learn soon that these clothes, especially the stiff dungarees, marked him as a State Boy, one of hundreds at Fernald with no family ties. (Their opposites were Home Boys, whose families brought them better clothes and took them away for occasional weekends and summer vacations.)

The humiliation of being undressed in this strange place made Freddie angry enough to drop his attachment to the state lady. He told her he didn't need her anymore. As the social worker departed, Nurse Jacques brought him into her office, where he sat on a chair and listened to her tap the keys of a typewriter, making entries in what eventually became a very thick file. She gave him the patient identification number 10130. Copying from the papers brought by the social worker, Jacques filled in Freddie's name, date of birth, place of birth and his parents' names. She noted that he was a Roman Catholic and that he was to receive free clothing.

In the medical section of Freddie's admission papers, Nurse Jacques wrote that he was in good health, and that the only distinguishing mark on his body was a "scar on right cheek from an accident coasting." Under the heading "Behavior" she reported that he "seems a very quiet boy." In the family medical history section, she noted "mother promiscuous," and that she was "colored Portuguese." As a child, the records showed, Mina had been case number 15482 within the state's Division of Child Guardianship. This meant that she was likely a foster child, too. For Freddie's father, Jacques recorded two stays in Boston Psychopathic Hospital—another state-run facility—and that he had been diagnosed with "mental deficiency" prior to his death by "carbon monoxide gas poisoning."

The most important entry box on the Fernald admission form was on the front page, a few lines below the spot where a photo was attached with a paper clip. It asked for Nurse Jacques to record a "clinical diagnosis." This entry was the legal and medical justification for Freddie's assignment to the institution. There, Jacques typed, "Familial—Moron."

After she pulled the last page out of the typewriter, Nurse Jacques brought Freddie into the hallway. Together they climbed the wide staircase.

"You might as well do what you do best. Make the most of it, because this is going to be your home. If you listen to what you are told, and do it, you'll be all right."

On the second floor, Nurse Jacques led Freddie down the corridor to the glass-pane door that opened on Ward 1. It was a room the size of a tennis court, with a bare oak floor, pale green plaster walls, and a high ceiling crisscrossed by exposed wooden beams. The faded light of dusk—it was now almost five o'clock—came through huge windows on three sides of the room. Thirty-six iron frame beds were lined up in rows two feet apart.

For a place that was filled with little boys, it was strangely quiet. As Freddie walked in, he noticed the shapes of a few boys lying in their beds, covered with sheets. Others sat on large, straight-backed wooden benches that had been arranged in a semicircle near the door. No one spoke to him.

Freddie could tell that the boys were roughly his age. Most wore the same clothes and big heavy shoes that he now wore. A few looked strange. They had large heads, prominent brows, and wide-set eyes. Others made noises that gave them away as mentally retarded. But most seemed to be just ordinary boys.

Once she had announced to the room that the new boy's name was Freddie, Nurse Jacques departed. A woman attendant who was the only adult in the room told Freddie to sit down and be quiet. The other boys eyed him carefully, hoping to see whether he was tough or weak, smart or slow. But Freddie didn't give them much to go on. He was so afraid that he couldn't speak or even look them in the eye.

Freddie squeezed onto a bench—at Fernald they were called

"settees"—and watched Nurse Jacques depart. The boy next to him, smelling of urine, mumbled the same syllables—"hulla-ma-shave, hulla-ma-shave"—over and over. When the attendant shouted at him to shut up, he did. This is how it went for the next hour or so. Every few minutes someone touched someone else and a word would be said. Once or twice a little laughter sneaked out of a boy's mouth. The attendant would shout "Shut up!" and the silence would be restored. Through all this, Freddie studied the faces of the other boys and tried to resist the impulse to squirm. It was a struggle.[2]

At five o'clock, the attendant ordered the boys to stand in a line, hold hands, and march to dinner in a downstairs cafeteria, where "waiter boys" served the meal from big bowls they carried from table to table. The smell of the food made Freddie feel nauseated. He couldn't eat. When the others finished, they returned to Ward 1. More boring, uncomfortable time on the settees was followed by a trip to the bathroom (the attendant called it "the Annex"), a tiled room with urinals, two shower heads, a steel sink with several faucets, and toilets without stalls. Fred went in with a group of boys who took their turns relieving themselves and washing their hands and faces.

From the bathroom, the boys went to a clothes room, where they changed into nightshirts and put their day clothes into numbered cubbyholes. Once they were in bed, the attendant ordered them to turn onto their right sides, facing the fire escape, so that no boy was able to look into another's face. The attendant snapped on the radio, which sat on a shelf near the door. Classical music filled the room. Then she turned out the glass-domed ceiling lights and went through the glass-pane door to sit at a desk in the hallway.

In the darkness, Freddie cried. No one had told him exactly where he was, why he had been put there, or how long he would stay. He imagined the place was a prison for boys, but he couldn't recall committing any crime that would have landed him in such a place. He glanced around the room, but with all the faces turned away, he was unable to connect with a soul. He heard some of the boys hum or repeat syllables over and over. "Hulla-ma-shave, hulla-ma-shave." Through the window, Freddie could see the moon glowing in space. To occupy his mind, he tried to think about the stories he had heard in

foster homes. The only one that came to mind was from the Bible—
Jonah and the whale.[3]

In his first days at Fernald, Freddie's immediate challenge was to adjust
to the routine of the Boys Dormitory, which everyone called simply "the
BD." Here nothing happened unless an attendant ordered it to happen.
In the mornings, an attendant flicked on the lights and told the boys in
the ward to wake up. Some, like Trudy Blacksmith, then walked down
the rows of beds grabbing at blankets and flinging them back. The few
boys whom she left alone were so disabled by cerebral palsy, muscular
dystrophy, or other ailments that they were unable to move on their
own and would be fed later. For some months, Freddie wondered if he
was destined to become bent and immobile like them, as if that were the
fate of every boy at Fernald. It was a terrifying thought.[4]

Once out of bed, the boys were herded to the annex in groups of
five or six. From there, they went to their cubbies, where they got their
day clothes. After they dressed, the boys were supposed to make their
beds. In each ward of thirty-six, there were a few who were disabled to
the point where this chore was too much. Brighter boys were assigned
to make those beds. Throughout the day, they would perform similar
tasks, almost like junior attendants, patiently aiding a child as he
dressed, or walked, or tried to feed himself.

It could take an hour for thirty-plus boys to pee, splash water, get
dressed, and make their beds. As they finished, the boys took their
places on the settees. On his first morning at Fernald, a tiny boy with
bowed legs, a brown crew cut, and yellowing teeth sat next to Freddie.

"I'm Albert," he whispered. He pointed across the room to a bigger
boy who sat on another settee and added, "That's my brother, Robert."

At first, Freddie was unable to reply with anything more than a
smile. Then he managed to ask, "What is this place?" Albert told him
he was in "the state school" and that as long as he did what he was told,
he would be all right. "Just do what I do."

When every bed was made and the bathroom was empty, the boys
were told to form two lines, hold hands, and follow an attendant
downstairs to the dining hall. Every day, it was the same: hot cereal,
toast, and milk. As tasteless as these breakfasts were, things were even

worse when it was over and they returned to the ward for endless hours on the settees.[5]

On three mornings each week, Freddie was freed from the settees to go to the Sense Training Center in Waverley Hall, the building closest to the Boys Dorm. When it was his turn, an attendant escorted Freddie and a few others downstairs and outside. In the minutes that it took for them to cross the courtyard, they were able to look at the Girls Dormitory where, sometimes, they saw girls their age coming and going. If it was warm, and the windows were opened, they might also hear the voices of teachers and children rising from the schoolhouse on the east side of the square.

The two upper floors of Waverley Hall were a dormitory for teachers, who received subsidized housing to make up for their low pay. Downstairs, one room on the first floor was used for the sense training. The rest of the floor housed severely disabled older men. Many of them were heavily sedated. Some tended to strip off their clothes just as soon as they were put on.

During the hour-long training sessions, which were based on the work of a nineteenth-century psychologist named Edouard Seguin, a teacher coaxed Freddie and the other boys through simple tasks, such as stringing spools onto a plastic cord, or weaving yarn through holes punched in cardboard. They were told to march in place, climb steps, or to practice tying their shoes. The boys enjoyed the training. It seemed like a game, and the teachers, mainly younger women, spoke softly, encouraged them, and might even hug them.[6]

Freddie performed well. The very first entry in his school record, written by Principal Mildred Brazier, noted that he "seems good material for training." The same report noted that when he was tested after his admission to Fernald, Freddie's IQ had gone up to 75, a full ten points higher than his results at Wrentham a year earlier. But it was still in the range for feeblemindedness.[7]

Given that he had bounced around in foster care since he was months old and had never been educated, Freddie's true IQ must have been even higher than the 75 he scored on his second Stanford-Binet exam. The staff at Fernald should have understood this, because they were aware of the shortcomings in the Stanford-Binet Test. Two years prior to Freddie's admission, a Fernald psychologist had written about flaws in the test in a report to Superintendent Malcolm Farrell. Nevertheless, attendants, teachers, doctors, and administrators continued to act as if a child's IQ score was reliable, and defining. It was used as if it were completely accurate, and was cited to support decisions about how a child would be educated or trained and where he might live.[8]

At the Boys Dormitory, more than a dozen relatively normal boys, whose IQ scores varied widely, lived in each ward. Invariably, these boys formed a sort of community with a certain hierarchy based on age, physical size, toughness, and intelligence. If you were one of them, you worked hard to make friends who would stand by you. Freddie first bonded with two boys who had already been there for a year: Albert, who had joined him on the settee when he had arrived in the ward and offered the first words of friendship and advice, and his brother Robert.

The Gagne brothers—on admission Albert was ten and Robert was nine—were the middle sons of a timid mother named Ruth and a brawling alcoholic father named Thomas. Mr. Gagne possessed a long criminal record and a small number of amateur boxing titles. According to state records, their mother, Ruth, "was said to be low-grade mentally, slovenly, lazy."

"Slovenly" hardly described the condition of the Gagnes' unpainted, wood-frame home in the coastal city of Newburyport. From the outside, the unpainted house appeared to be abandoned. Inside, grease and dirt covered everything. The coal stoves that were supposed to heat the place were often cold in winter for lack of fuel. Water pipes froze, and when toilets didn't work, the family just defecated and urinated on the floor. Lice and fleas infested the single bed where five children slept. Rats ran freely.

On many nights, the Gagne children would lie in the dark and listen to their father beat and rape their mother. "We didn't know exactly what was happening, just that he was beating her and she hated it," recalled Doris (Gagne) Perugini, older sister of Albert and Robert. Sometimes the children hugged each other. Sometimes they rocked rhythmically, trying to soothe themselves with motion. During the days, the children were practically feral, roaming the town begging for coins and stealing from shops. They were so dirty and sick from scabies, rickets, and infections that they were shunned by neighbors and sent home by the local school.

The Gagne children became known to state authorities in February 1944, when the kitchen stove went cold. The children went out with their mother, who pushed a barrel loaded onto a baby carriage. They stopped at a coal company yard. She instructed the children to scramble under a fence and come back with lumps of coal to throw into the barrel. A watchman saw them and called the police. Ruth was arrested and taken, children and all, to jail.

Doris would long remember that her mother cried, the tears cutting a track through the dirt on her cheeks, as a municipal court judge admonished her to take better care of her children. Days later, on February 18, 1944, state social workers appeared at the Gagne home and took four of the five children—the eldest boy, Wilfred, was not home—into custody. Albert and Robert were separated from their siblings and placed with a foster family.

Like Freddie Boyce, the Gagne boys lived in a series of temporary homes. The last was in Peabody. The brothers would recall the woman who ran the house as violent but also indulgent. In the evening, she had the boys brush her waist-length hair while she talked to them. In the morning, she hung their soaked sheets out the window to humiliate them when they wet their beds. Perpetually frightened, lonely, and plagued by nightmares, the boys wet their beds quite often.

The home in Peabody was so awful that Robert was actually happy the day that a judge sent him to Fernald. At the courthouse, he had been allowed to fill his stomach with peanut butter sandwiches and milk. That lunch was one of the best meals of his life, one of the rare moments of his early childhood when he wasn't hungry. When he got to Fernald, Robert appreciated the orderly, predictable environment. His records show he got a hot bath within a half-hour of arriving at the Boys Dorm and was given a fresh set of clothes. Though a boy who had known a better life might have chafed against the regime at BD, it was a comfort to one like Robert.

Albert Gagne, who had soft gray eyes and light brown, almost blond hair, had more trouble adjusting to life in an institution. His bond with his siblings and parents was stronger than Robert's, and he missed them. He was also subjected to far more teasing and bullying in the ward than his brother. Malnutrition and childhood diseases, including rickets, had left him grossly undersized and knock-kneed. At age ten, he was just four-and-a-half feet tall and weighed less than sixty pounds. His teeth were yellowed and rotting. He clung to his brother, and cried frequently. Robert, eight inches taller than his older brother, reminded Albert that they were together, at least, and that they could take care of each other. Perhaps it was Robert whom Malcolm Farrell had in mind when he wrote to the boys' parents on the day after their admission. It would be the only time anyone at the school would report to them on the condition of their children.[9]

> Dear Mr. and Mrs. Gagne:
> I am writing to inform you that your sons, Albert and Robert, were admitted to this school yesterday and placed in our Boys Dormitory, where they have shown no homesickness and seem to be making a very satisfactory adjustment to their surroundings. They will be placed in school classes and trained to the limit of their mental ability.
> I suggest that you not visit for a month to give the boys a chance to adjust to us.
> I will not write again unless they become acutely ill, but would be very glad to answer any questions you may care to ask.
>
> Very truly yours,
>
> MALCOLM J. FARRELL, M.D.
> Superintendent

While Malcolm Farrell assured Thomas and Ruth Gagne that their sons had immediately adapted to life in the institution, in truth the boys struggled to find a place in the ward's social order. Though they preferred to keep each other company, inevitably other boys intruded, often to pick on Albert. Freckle-faced, with brown eyes and dark hair, Robert adopted the role of "big" brother, the protector who defended Albert when other boys shouted, "Corn-teeth! Bow-legged!" and "Midget!"

Earl Badgett* and Billy Mason, two of the tough boys whom the others called "Big Shots," taunted Albert Gagne almost every day. Badgett was thickly built and strong. Mason—tall, thin, blond, and blue-eyed—was less imposing but more aggressive. Albert was afraid of them and wouldn't even respond when they called him names, or pushed him. Robert hated the way they went after his brother and worried about the possibility that he might have to fight both of them at the same time.

One morning, the two Big Shots came upon Albert alone in the bathroom, and their teasing echoed off the tile walls into the ward. Robert Gagne heard it and rushed to stand by his brother. Freddie turned to watch him and listened. He heard Robert tell the bigger boys to leave his brother alone. He heard Billy Mason call the Gagnes "lifers" because they never received visitors, and this meant that no one outside the institution would ever care enough to try to get them out.

The insult hurt because it was true. On every Company Sunday—a monthly visiting period that lasted just four hours—the Gagne boys were among those who waited and gazed out the window for visitors who never came. Downstairs, relatives of other boys arrived at the office and had the child they wished to see brought down to a common space called the "day room." Some were allowed to take a child out for a ride in the car, a visit to a candy store, or an afternoon at home. (When a boy returned from an outing, he was expected to share whatever gifts he had received, whether it was candy or clothing. Sometimes Big Shots would simply take what they wanted from the more timid boys.)

Every boy who was left alone upstairs on Company Sunday was jealous of those who were visited. For these boys, including the Gagnes, "lifer" hit too close to their sense of hopelessness. Freddie, who never received visitors, felt it. He knew how ashamed and disappointed they felt about being abandoned, possibly for life. This is why he acted when the Big Shots began calling Albert a lifer.

"Shut up!" shouted Freddie as he ran into the bathroom where the Big Shots teased Albert Gagne. He balled his fists and stood beside Robert, evening up the sides. Earl turned, spat into the sink, and muttered that Albert wasn't worth fighting over. As he and Billy walked away, they repeated the curse—"lifer." For the rest of their time at Fernald, Robert and Albert were Freddie's friends. Though they sometimes had trouble understanding him because of his speech problems, this was not a real obstacle. With no attentive adults to teach and correct them, many of the boys had garbled, slurred, and quirky speech patterns. Sometimes a single boy's erroneous choice of a word or mispronunciation—"He's a 'baffoon'"—was adopted by everyone else. In other instances, the malapropism came from a staff member—"I 'pacifically' told you to sit down!"—and became universal on the ward.

Freddie appreciated the way Robert and Albert Gagne accepted him, with all his flaws. They would pretend to understand everything he said, going along until they figured what he was trying to express. They would never humiliate or deceive him, and their friendship was consistent. But as much as Freddie liked the Gagnes, his closest buddy would be a short, curly-haired boy named Joey Almeida.[10]

Unlike many of the boys who came from the foster care system, Joey Almeida was a defiant, almost delinquent kid. Prior to being committed to Fernald, he and his older brother Richie had lived with their stepmother in a shabby tenement apartment in a poor section of Cambridge. Their birth mother, who had served prison time for breaking into a jewelry store, had deserted them. Their father, who had immigrated to America from the Azores, was away for weeks at a time working with crews erecting fences along newly built interstate highways. The boys were too much for their stepmother to handle. She often beat them and then kept them out of school for fear that the marks she had made on their faces, arms, and bodies would be discovered.

Richie and Joey spent many of their days roaming the streets and stealing everything from cupcakes to bicycles. On summer nights, they sneaked out to wander through parks and doze like stray cats on rooftops where the black asphalt radiated the heat of the day. When they needed money, they wandered through a bar on Inman Square to beg nickels and dimes. When it rained, they broke into garages or crept onto someone's back porch to stay dry. When they had time to think, and talk, they fantasized about their father, a big man with a crew cut and a loud voice. They daydreamed about him rescuing them from their stepmother.

In reality, Anthony Almeida was unable to solve the problems of his sons and his second wife. After years of dealing with police, truant officers, and family court judges, he allowed the state to test his sons for mental deficiency. The testing, which failed to take into account their lack of education, years of parental abuse, and utterly chaotic lives, led doctors to report that both boys were borderline "morons." (Erroneous as it would seem years later, the label would stick and determine their fates.) Mr. Almeida agreed to commit them to Fernald, where, supposedly, they would receive appropriate care. As an important side benefit, Anthony Almeida would get peace and quiet at home.

On a warm Friday afternoon in early September 1949, Mr. Almeida summoned his boys to his car, told them to get in, and began to drive. He stopped at a store and bought two tins of peanut brittle, which he told his sons to leave unopened for "later." He steered the car west, through Cambridge, past McLean Hospital, the famous, private psychiatric institution that served Boston's elite. As the car passed through the gates of Fernald, Joey couldn't read the words on the sign, but Richie could. Years later, Joey guessed that it was the sign that made Richie cry as they got out of the car. He must have known what was happening.

Richie whimpered as he sat with his brother on the bench outside Nurse Jacques's office in the BD. Joey stayed calm while his father was inside the office talking. He whispered to Richie that he should be quiet. Richie couldn't stop. When their father emerged from the office, Richie called out, "Dad, Dad!" Mr. Almeida told the boys to calm down—he was just going outside to get something he had left in the car. They believed him for a few minutes, but when he didn't return, Joey began to suspect that he wasn't coming back at all.

When Nurse Jacques came out with state clothes, the Almeida boys finally realized exactly what was happening to them. They changed in the hallway. Nurse Jacques then brought them upstairs, took Joey to Ward 1, and tugging Richie's hand, led him to a separate ward, on another floor.

Joey still didn't cry. He knew he was supposed to stay in the ward with all the other boys, but he wasn't convinced that his father had abandoned him for good. Also, in the back of his mind, he knew that he and Richie had managed on their own in the outside world. If necessary, they could run away together. Until that could be arranged, he would do his best to adapt to his new home. But he vowed he would never accept that he belonged there. He was not a moron.

That night, as the boys in the ward washed and dressed for bed, they sized up the new kid. He was not especially large, but he looked tough and confident. For a child who had been through the ordeal of admission, he was especially calm. He seemed like a boy who could take almost anything and not back down.

For his part, Joey scanned the faces to see who looked like a Dope and who might be smart enough to qualify as an equal. Joey picked out Freddie as a potential ally and friend. He then went to the cubbyhole where the tin of peanut brittle was hidden among his clothes. He grabbed two jagged pieces, put one in his mouth and then walked back to the ward with the other hidden in his palm. As he passed Freddie's bed, he handed it to him without breaking stride. When he got back to his own bed, he turned to see Freddie's cheek bulging with the candy. After he swallowed, Freddie smiled. Joey smiled back.[11]

Over months that became years, the residents of the Boys Dormitory coalesced into a group with an informal hierarchy. First were the tough, dominating boys like Earl Badgett and Billy Mason—the Big Shots. Next came boys like Freddie and Joey, who could defend themselves in a fight but were also smart. Third were the more timid regular boys like the Gagne brothers, who sought safety by sticking together. Last were the so-called Dopes, more severely retarded boys who sometimes received the rage and anger of the Big Shots and couldn't fight back.

The roles were neither permanent nor unbending. On many occasions, a boy who was considered a weakling fought back and earned an end to harassment from the Big Shots. On other occasions, a group of boys would stand up for someone who was being bullied. Freddie often felt this impulse to protect others, especially when it came to the smaller boys, such as Albert Gagne and an even more vulnerable child named Robert Williams.

At age eight, Bobby Williams was just three feet eight inches tall and weighed only forty-six pounds. He had blue eyes and straight red-dish-brown hair. Two years before Freddie's arrival, he and his twin brother Richard had been placed in different wards of the school. This separation had caused them grief that Robert had yet to overcome.

The Williams boys had been born to an unwed mother who relin-quished them to the foster care system. Like Freddie, they had enjoyed the best moments of their lives in their last foster home. "In this home," explained a note in Bobby's state record, "the boys were greatly loved, caused no trouble and were very healthy."

George and Mary Yeo of Marlboro had treated the Williams twins like their own. When social workers suggested the boys go to an insti-tution, Mary insisted she had prepared them for school and could care for them long-term. "Foster mother at that time felt a great injustice had been done, and that the boys were really learning and were not stu-pid," notes Bobby's record. Mary Yeo's protests were futile. Once experts at Fernald had determined that the twins were both morons with IQs below 70, they were destined for institutional life.

At Fernald, Bobby was alone, without his twin, for the first time. Tougher boys harassed him. Three months into his stay, he was shoved off the toilet and broke an arm as he fell onto the tile floor. (The details of the incident were recorded by an attendant in her daily log.) The experience taught Bobby to be especially wary of certain boys, to bring a friend to the bathroom, and to yell for reinforcements when threat-ened by Big Shots.[12]

Unfortunately, none of the boys could call for reinforcements to help in a confrontation with the most formidable group inside the BD—the attendants. These adults ruled the wards with complete power and authority. Most were well-meaning and rarely became vio-lent. But a significant number of them were dangerous. The boys stud-ied them moment by moment. And when they were alone, they talked incessantly about them, trading clues to their behavior and venting their anger. They knew that the attendants were not supposed to beat and berate them. And in their fantasies, they ran away, grew up, and then came back to avenge themselves. Some imagined coming back to attack Lois Derosier, who had hurled the bowl of urine at Howie. Others had it out for James McGinn.

"McGinn was always annoyed that Ms. Derosier could get us to be quieter," recalled Bobby Williams decades later. "She did it by sneaking up on you, surprising you when she hit you. It could be a slap, or she could bang your head against the back of the settee, because you had been loud, or laughed. You were always looking for her, and that kept you quiet. McGinn did things in a way that you knew what was coming. I think that's why some kids pushed him more. They felt like they could tell when to stop before he hurt them. But that didn't always happen, and the stuff he did made you feel really bad."

A tight little man with short black hair, a straight back and sharp eyes, James McGinn always looked like he was annoyed and poised to attack. He couldn't tolerate noise. When he came on duty, clad in his pressed white uniform, he jingled his keys as he walked down the hallway toward the ward, a signal to the boys that they better stop talking. If McGinn entered the room and didn't like what he heard, he was likely to use his keys, which were fastened to a long cord, to strike whoever was talking.

Unlike other violent attendants who were unpredictable and might suddenly become enraged, McGinn was methodical. For example, during almost every meal in the crowded first-floor dining room, he walked silently around the tables with a large metal spoon in his hand. A boy who talked too loudly, or talked at all when McGinn had commanded silence, would get a quick whack on the head with the spoon. At other times, McGinn declared that he was searching for "hollow heads" and worked his way around the tables thunking as many skulls as he pleased.

Up in the wards, where he was often alone with the boys, McGinn allowed himself to practice sadism with a sexual component. Instead of red cherries, he would have the boys line up, pull down their pants, and stand still while he went down the line yanking each one's testicles. The pain was almost equaled by the shame the boys felt as they trembled with fearful anticipation. Many began crying as McGinn approached, well before he even touched them.

Any boy with real intelligence could see that you needed to be especially alert when McGinn was on duty. You never turned your back on

McGinn, and you walked away if somebody started talking loud, laughing, or acting foolish. Smart boys noticed that some of the others could forget where they were, and who was on duty. They would whisper a reminder: "Cut it out! It's McGinn today."

Freddie could avoid McGinn for months. But one night, he couldn't resist shouting "Stop" when the attendant raised his wooden spoon over Robert Williams's head. McGinn decided to hit Freddie instead. As he whipped the spoon against Freddie's head, he grunted the word "my." He struck again and said, "black-assed." With the third strike he said, "little," and with the fourth he spat out, "nigger boy." This phrase—"my black-assed little nigger boy"—became his name for Freddie and a curse that other boys used when they wanted to make him feel bad. (There were just a handful of black residents at Fernald, and none in Freddie's ward at the time.) [13]

When his name-calling and routine of abuse failed to keep the boys in line, McGinn employed a number of inventive methods to inflict pain and dread without leaving any marks. The first time Freddie saw this brand of discipline, a couple of nine-year-olds could not stop talking as they came back upstairs from lunch. McGinn pulled the two over to their beds and ordered them to strip and kneel on the iron frames that jutted out from under the mattresses. They were to stay there, with the metal pressing against their tendons, until McGinn allowed them off.

With the two naked boys struggling to keep their balance, McGinn turned to the others and announced it was time to clean, wax, and polish the oak floor. He organized some of them to push the beds aside and sent others to a supply closet for the wax and rags. A third group got on their knees to start digging dirt out of the narrow spaces between the oak planks. The grit they dug out was collected with the rest of the dust and dirt when the floor was swept.

When the floor was clean, and the supplies were at hand, McGinn had some of the boys take pieces of wax, which came in blocks the size of bar soap. He ordered them onto their knees, and they began rubbing the white blocks across the floor, leaving smears of wax with each stroke.

Once the first line of boys had worked a few feet down the floor, another group equipped with soft rags followed. The rag boys rubbed as hard as they could, to make the smears made by the wax boys disappear into a glossy shine.

But a mere shine was not enough for McGinn. To make the oak shimmer like new brass, it needed what was called "the rope rub." McGinn brought out a rope and a four-foot-long wooden beam covered in carpet. The rope was passed through big rings that were screwed into the ends of the beam. Two boys grabbed the ends of the rope, and a third sat on the beam as they dragged it across the floor over and over again.

Perhaps ten minutes passed, with small hands waxing and polishing as fiercely as they could and the beam dragging heavily across the floor, when the first kneeling boy began to cry. Five minutes later, he fell off the bed frame and, unable to walk, lay there. The second boy lasted a few minutes more, but though he never cried, he fell, too, and remained where he was, with his knees pulled up to his chest.

It took more than two hours for the waxing to be completed on half the room. All the while, McGinn warned the boys that they would be next on the bed irons. If they didn't like that punishment, he said, they could go to Ward 22, an isolation unit in a building halfway across the campus. (None of them had ever been to the punishment ward, but the threat was fearsome. They had heard that 22, as everyone called it, was a dangerous place, where boys were restrained in straightjackets and tied to their beds with thick leather straps. This terrible reputation meant that the threat of being sent to 22 was enough to silence any ward in the BD.)

Once one side of the floor was polished, McGinn organized the boys to move the beds. He barked for the two who had been punished on the bed frames to rise and help. McGinn threw rags at them, and they fell in with the polishers as the second half of the floor was slowly given a shine.

The bed-iron treatment was a common punishment, and the floor-polishing routine, including the rope rub, was a duty the boys were required to perform several times each month. At various times the floors of the hallways, and the stairs and landings, were added to the chore. "Sometimes you'd have to get down with a butter knife and dig dirt out of the cracks," recalled Joey Almeida decades later. "But it was something to do. It broke the monotony. That was good."

Many former State Boys would express similar feelings as adults. They would recall cracking jokes as they labored over oak floors, doing anything they could to make daily life more bearable. Strange as it may seem to an outsider, many of the boys in the BD found ways to accept what happened to them in the wards, and even tried to understand the behavior of attendants like McGinn. It was one way that they could make sense out of their world and perhaps even learn to predict what might happen next.

As an adult, Fred would be able to put into words the impressions he developed as a child. "I could see the attendants were overwhelmed," he said, "and a lot of them weren't that smart, or patient, to start with. When things were bad, the noise in the ward could be deafening, with all these kids talking and making noise. It was a real cacophony. And a lot of the time, one attendant would have to handle a ward alone, or two wards. It was practically impossible. They would lose it."

Being an attendant at the Fernald School meant keeping dozens and dozens of residents safe, clean, and sheltered for eight hours straight. In some dorms, the charges were adults who could not talk, dress, or use the bathroom. In others, like BD, they were young, energetic, and difficult to control. The challenge, in either case, was more than most people could handle, especially those who were sympathetic and refused to resort to violence or threats. The burden was so extreme that each month brought a dozen or more resignations. It was not unusual for a new employee to simply walk away from Fernald after a week, a day, even an hour on the job. The school was constantly hiring.[14]

In the month that Freddie Boyce became a State Boy, a man named Sumner Noble joined the staff caring for the males at Fernald School. (Only women were allowed to oversee female residents.) A Coast Guard veteran of World War II, Noble had found it difficult to land a job in a postwar economy flooded with former servicemen. He lived near the Fernald School and followed a sister, who had gone to work there first, into a job as an attendant. With no seniority, Noble was not able to get a steady posting. Instead, he was shifted from building to building, and worked all different hours.

Wherever Noble was sent at Fernald, he encountered something that made him feel like quitting. In the North Building, behind the Boys Dormitory, he found wards where dozens of severely disabled men—some of them stark naked—wandered aimlessly in large, cold, unfurnished rooms. Many were drugged to keep them quiet. Others babbled incoherently or rhythmically pounded the walls and floors with their feet, hands, even their heads. Puddles of urine and smears of feces made the floors almost impassable. Attendants, at most two for a ward of thirty-six, were barely able to keep the men from harming themselves and each other.

The Boys Dorm and the nearby Boys Home, which housed high-functioning adolescents, presented their own depressing realities for a new employee. In BD some of the little ones clung to the attendants, begging for affection. Others repeatedly asked for mothers and fathers who never came to visit. Pushing these needy children away required physical strength and left a sensitive person feeling heartsick. At the Boys Home, tougher, emotionally withdrawn adolescents demanded to know why they were being held prisoner and how they could get out. In both places, Noble saw boys who were no more retarded than he was.

"I understood that the really bad ones could be sent to the state's farm at Templeton Colony, where they might stay the rest of their lives, so we would warn them about that. And there were a few, very few, who were released from time to time. But I didn't know how that happened," says Noble, "and couldn't give them the answer they wanted."

Noble recalled his time at Fernald from the age of ninety-four, but as he sat on the porch of his home deep in the Maine woods, he insisted his memory was clear. "I remember thinking that a criminal in state prison at least knew when his time was up and he could look forward to getting out. These poor little devils didn't even have that. And you better believe they were angry about it."

Noble recognized that the boys were powerless. Any complaint, stacked against an attendant's word, ended with supervisors siding with the attendant. Speaking up was no use. The same was true for staff at the school. Supervisors didn't want to hear about problems, and the more violent attendants were actually favored because they got the job done. So Noble kept what he saw, and what he heard, to himself. And like so many others, he quit in a matter of months. He found a new job at a machine shop and worked part-time as a police officer in Waltham. However, the horrible memories of Fernald never disappeared from his mind.

"I knew what was going on there. You couldn't miss it," he says, knitting his frail fingers together on his lap. "There were no jobs around. I needed the work. All through life, I've been the kind of guy who didn't talk about things. But a place like that, even if you don't talk about it, you never forget it."[15]

THREE

Fernald, like all large institutions, was a world apart, with its own rules and reality. Though many attendants were not violent, they were all gods in the way that they controlled the environment. The boys depended on them for food, shelter, physical safety, and for higher-order needs such as dignity and hope. Their complete power, and the obedience it fostered, guaranteed that the boys in the wards would not speak out or resist when something terrible occurred.

The State Boys of BD were among the youngest and weakest citizens in this bizarre community, and attendants routinely preyed on their ignorance and imaginations to make them obedient. They continually threatened to give the boys shock treatments or lobotomies "so you come back like a zombie." Exile to Ward 22 was often raised as a possibility, as was transfer to "the loony bin at Met State." (The boys knew that Metropolitan State Hospital, a massive psychiatric institution equipped for shock therapy, lurked a few miles away. They were not aware, however, that the more genteel McLean Hospital, a premier psychiatric facility for the rich, was also within a mile of Fernald. At the time when Fred and his peers occupied the BD, the soon-to-be-famous Sylvia Plath was at McLean, receiving electroshock.)[1]

In response to their fear and anxiety, many boys retreated into a fog of dissociation and docile obedience. In this state, they viewed beatings, torture, and sexual assaults as if they were watching a film. Moments later, it would be almost forgotten. Freddie Boyce came to call the feeling of detachment that he experienced at the BD "the ether." Other State Boys said that it affected Freddie more than others.

"He seemed like he was off on a cloud," recalled Robert Gagne, "or sometimes he would have what people nowadays call anxiety attacks. He'd say his heart was pounding and he would get up and walk around at night."

If the ether wasn't enough to obscure what certain attendants did—including physical, emotional, and sexual abuse—then they could rely on the fact that they were often completely alone with the boys they controlled. Visitors were barred from the wards, and some residents could go for years without encountering an adult who might consider their complaints. Even then, who would believe the word of a supposedly retarded boy?

Similarly, the boys of Fernald could not count on the administration of the school for help. Superintendent Malcolm Farrell, a formal man with a military background, was so rarely present in the wards that many residents didn't know what he looked like. Farrell was a "big picture" bureaucrat who excelled at budgeting, public relations, and politicking with state officials. He prized routine and efficiency and often said that he didn't need to be informed about how a staff member solved a problem "as long as it's legal." Their effectiveness allowed Farrell and other top-level professionals at the school to focus on important matters of science and policy. Though they may have seen residents on the grounds, and dealt with various crises, for the most part Fernald was a backdrop for their careers, the source of their authority in the outside world, and a handy reservoir of human material for research. One sign of how they regarded the residents was their use of the word "item" to refer to individuals in the wards.[2]

It would have been natural for a man like Farrell to overlook the daily chore of caring for Fernald's 1,900 residents. The post he held as the superintendent of the nation's oldest state school automatically made him one of the most prominent psychiatrists in the country and a recognized expert on the care of the feebleminded. Every year, he welcomed scores of visitors from other institutions who sought answers to their problems and a model to follow. Farrell held several important posts in the American Psychiatric Association, and he was in demand as a public speaker who could win support for the cause of mental health.[3]

The postwar years offered opportunities that inevitably drew Farrell's attention away from the wards in his own school. America had begun a Cold War with the Soviet Union, and science was one of the battlefields. Federal agencies were making hundreds of millions of dollars available for all kinds of research. Some of this money was reserved for institutions like Fernald, which eventually allowed outsiders to use its residents as research subjects for studies on everything from neurological development to nutrition.[4]

Farrell and others who ran the nation's schools for the feebleminded had long sought to elevate their institutions from mere asylums to centers of science. Henry Goddard, author of *The Kallikak Family*, founded a research lab at the New Jersey Training School in 1906. (Research subjects included metabolism, psychology, heredity, and the study of head measurement.) And as early as 1915, doctors at Fernald teamed with colleagues at Harvard Medical School to study the heritability of feeblemindedness. Over the years, investigators at Fernald followed the changing theories of the field. After genetic causes were investigated, their focus shifted to endocrine disorders.[5]

In 1949, the leading proponent of the idea that malformed or malfunctioning glands were responsible for feeblemindedness was Clemens E. Benda, the psychiatrist and neuropathologist who headed Fernald's on-campus medical laboratory. A mild-looking man with pale skin, receding hair, and a fondness for bow ties, Benda had emigrated from Germany in 1936. He was the son of a famous pathologist and grandson of a noted theologian. As a student and young doctor, Benda had worked under both Karl Jaspers and Ludwig Binswanger, pioneers in existential psychology. In America, he became a research fellow at Harvard and a lecturer at Tufts before he was hired, in 1947, at Fernald.

Benda suspected that nearly all forms of feeblemindedness were caused by a malfunctioning pituitary, thyroid, or other gland that disrupted the brain's normal development. The problem occurred either in utero or during early childhood, he thought. He sought proof of his theory in the brains of those who died at Fernald and other institutions, which he examined, millimeter by millimeter, with the hope of finding telltale differences in size and structure.

Brains, as well as nerves and spinal cords, were readily available at Fernald because deaths from accidents, pneumonia, and tuberculosis were common. During Freddie Boyce's first month at the school, two inmates died of pneumonia. In the summer that followed, children as young as five and adults as old as seventy would die from a variety of causes. One woman's death from pneumonia was investigated by the local medical examiner because her autopsy revealed a black eye, bruises, and broken ribs. Nothing came of the probe, and later Benda reported preserving her brain and pituitary.

Working in the basement of a squat brick building called the Southard Laboratory, Benda accumulated a collection of more than 50,000 samples of brain tissue. For a while, he was so certain that these proved the thyroid's role in feeblemindedness that he wrote that an outright cure might come from his work. One of his experimental treatments involved feeding children with Down syndrome thyroid and pituitary hormones. Although the treatment raised hopes and brought many eager families to Benda, it would prove ineffective over the long term.[6]

Outside of his own lab, Benda tried to establish a reputation as both a scientist and public figure. After a visit to Germany in 1949, where he benefited from being a native speaker, Benda assessed the condition of children who had lived through the war for the *Boston Globe*. In this article, he reviewed the kinds of trauma suffered by children in war, and then noted among them a great increase in "stammering and tantrums" as well as stunted growth due to malnutrition.[7]

For a laboratory scientist, Benda got around. In 1949, he also went to Montreal and Paris for conferences and to Baltimore to lecture at Johns Hopkins Medical School. He even went to Washington to observe Dr. Walter Freeman performing a lobotomy, which Benda reported "may have some possibilities in an institution for the feebleminded in the future."

Freeman was America's leading lobotomist and performed roughly 3,500 of these operations, including one on John F. Kennedy's sister Rosemary. Eventually, lobotomy would fall out of favor as terrible side effects and doubts about its benefits were recognized. But at the time that Benda visited Freeman, the practice was at its height. More than

40,000 operations would be done between 1945 and 1955. And in 1949, the doctor who invented the treatment, Antonio Egas Moniz, was rewarded with a Nobel Prize.[8]

Benda's travels and exploration of psychosurgery reflected the range of his scientific interests and his desire to raise Fernald's status as a center of science. In his first ten years on the job at the school, he would build a research staff of twenty-four, increase grants tenfold, and welcome a large number of collaborators for studies. He became best known for a study of nutrition that had nothing to do with the Fernald School's mission but brought him into direct contact with the State Boys, including Freddie Boyce, the Gagne brothers, and Joey Almeida.[9]

In postwar America, many researchers turned to state institutions for human subjects to use in experiments. Perhaps the most famous of these studies, done in the same time period as Benda's work on nutrition, involved early polio vaccines. Residents of the Sonoma State Home in California and of the New Jersey State Colony for Feebleminded Men in Woodbine received the first test doses of the serum developed by Albert Sabin. Residents of the Polk State School in Pennsylvania got Jonas Salk's vaccine. Some risk was involved, since earlier polio trials had left one child dead and others paralyzed, but fortunately no injuries occurred.[10]

Compared with what Salk and Sabin were up to, the Benda-MIT project was minor. This was reflected in the note that Malcolm Farrell sent to parents in autumn 1949, explaining the research:

> The Massachusetts Institute of Technology and this institution are very much interested in the various aspects of nutrition, particularly how the body absorbs various cereals, iron and vitamins. We are considering the selection of a group of our brighter patients including _____ to receive a special diet rich in the above mentioned substances for a period of time.

Farrell sought parents' permission to enroll boys from the BD in the study, which would include the "special diet," frequent blood tests, and the collection of urine and feces. The information gathered would be of "considerable benefit to mankind," he wrote. Permission was granted for nearly every boy whose parents were approached. In cases where a guardian could not be found, the state assumed responsibility and enrolled him in the experiment.[11]

To encourage cooperation, Dr. Benda and his MIT collaborators began calling the boys they selected—Freddie, Joey, the Gagne brothers and others—the Science Club. They were told that they were special, the smartest boys at Fernald, and that they were making a contribution to society. They would be rewarded for being part of the experiment. Trips to ball games, parties, even gifts were mentioned.

Anything that broke the sameness of daily life at Fernald held enormous appeal for the boys. The idea of actually leaving the grounds to watch Ted Williams and the Red Sox was extremely exciting to them. (The team had barely missed winning the pennant in the previous season and was the talk of New England.) But more significant than these rewards was the status that the Science Club conferred. It was for the smarter boys. They were proud to belong.

In reality, while Dr. Benda wanted boys who were bright enough to behave and follow directions, he was mainly looking for those whose height, weight, and overall health were typical for American boys of certain ages. After the selections were made, the researchers discovered that Science Club members were not quite perfect subjects. They were malnourished, and had to be fattened up with an extra pint of whole milk per day, per boy, for a full month.

Once they were ready, the Science Club boys were moved into the day room on the first floor of the Boys Dormitory, where they were to be quarantined for a month. Benda and his staff could make sure they didn't eat anything that wasn't prepared and delivered as part of the study. This food, as well as their waste, would be measured and weighed.

The precise purpose of the study, which was funded in part by the Quaker Oats Company, was to see if naturally occurring chemicals in oatmeal called phytates affected the body's absorption of the calcium in milk. The protocol required the Science Club members to eat meals of oatmeal or farina wheat cereal. In some cases, the cereals were pure. For other meals, they were combined with the phytates. Lab technicians would compare the calcium they ingested with the amount they shed in urine and feces. The boys would also give blood samples, which would be tested for calcium levels.

The boys didn't know the details, but they could see that the scientists were exceedingly careful. They donned rubber gloves and white gowns to prepare the food. After it was served, they hovered around the tables, noting the time each boy started eating and the time when the bowl was finally cleaned. It wasn't always easy to eat the stuff they served. The portions were big, and the taste was boring, but the boys took pride in being junior scientists. For once, they were valued by intelligent adults and treated as if they were important.

Any other group of children might have balked at the services required after each meal, but the members of the Fernald Science Club willingly filled jars with urine and used a special commode that captured every bowel movement. (Those who found the second step a challenge were helped with a dose of laxative.) The blood tests involved a little pain, but it was a matter of pride to submit without crying. The more stoic boys felt they had really earned the rewards that came when the trials were over.

The first big reward for the Science Club boys was a visit to MIT on a cool December day. The boys all piled into a yellow school bus, which rolled to Cambridge, onto the campus, and then stopped in front of the Faculty Club. Inside, a Christmas tree and an elaborate turkey dinner—no oatmeal or farina—awaited the boys. They ate, and then listened to various speakers thank them for their contributions to science. Finally, Santa Claus appeared and gift boxes were handed out. As the boxes were ripped open, each boy found the same treasure—a Mickey Mouse wristwatch.

The watches, party, and praise from adults were deeply symbolic for the Science Club boys. They were special, at least among the residents of Fernald, and their lives mattered. They all wanted to be recognized as individuals, to achieve something distinctive. On the outside, boys their age would have done this by winning awards in school, as scouts, or in sports. At Fernald, they could do it as research subjects.

The Science Club would come together several more times before the completion of the research in 1955. Each study went more or less like the first, with the State Boys moving into a day room and eating a special diet. At a second Christmas party, the gift boxes contained Hopalong Cassidy mugs.[12]

Whenever rewards were offered at Fernald, manipulative attendants could exploit them. A State Boy named Larry Nutt would forever recall how he paid an extra price before he was allowed to go on a promised Science Club outing to Fenway. On the day before the trip, an attendant confronted him on a path that cut through some woods separating two campus buildings. The attendant, who was accompanied by two Big Shots, pulled Larry into the trees and demanded oral sex.

"He knew that I was on the list to go to the game the next day," recalled Nutt many years later. "If I didn't do it, he would tell them to leave me behind."

While the Big Shots stood watch, Nutt kneeled and then gagged and cried as he did what he was told. Though the assault was over in a few minutes, it felt much longer to Nutt, who heard the other boys laughing as the attendant gripped his head and rocked back and forth. As an adult, Nutt could not remember the ball game, but he recalled everything about the incident in the woods, right down to the sounds made by the attendant, the smell of his body, and the dirt that was on his knees when he got up. It would affect him for the rest of his life, filling him with sexual shame and rage.

Given how much the State Boys hungered for a break from the isolation of Fernald, it is possible to understand how Larry Nutt could be pressured to perform a sex act in exchange for a field trip. Fortunately, other attendants at the school offered some of the boys outings without demanding anything in return. These were rare gifts, and they soaked in the experience like tourists in an exotic land.

On one winter weekend, an attendant who had noticed that Robert Gagne never had a visitor got permission to bring him to his home for an afternoon. Every moment in this outing made a big impression on Robert. He hadn't been inside an automobile more than a few times in his life. The sound of the engine, the feeling of acceleration, the way a sharp turn threw his body from side to side—it was scary and fun. But an even bigger thrill awaited him at the attendant's home. Though television was fast becoming a staple of life in America, this was the first time Robert had ever seen it. He sat close to the set in the attendant's living room and tried hard to commit what he saw to memory, so he could tell his brother about it.

Back at the BD, Robert told Albert, Freddie, and a few others about what he had seen. They were curious, and jealous of his experience. Some boys didn't believe Robert's description of television, but they would soon learn he was telling the truth. In 1950, some relatives of Fernald Home Boys donated televisions to the institution. The Boys Dormitory got one of the first sets, and it was installed in the day room on the first floor. Four wards—more than one hundred boys—had to share access to the TV, so typically each boy got one night of viewing per week.

Like all boys their age, Freddie and his peers liked most the action shows such as *Adventures of the Lone Ranger* and *Captain Video*, "Guardian of the Safety of the World." But they watched anything that came on. And they came to value the privilege of watching the television more than any other. The tiny screen was a window on the outside world.[13]

The "outside" began to come to the State Boys in ways other than TV. Beginning in the late 1940s, civic organizations and a handful of individuals began to take an interest in the place. One of the first was Bea Katz, a thirty-eight-year-old homemaker who lived in a suburb near Waltham and belonged to a synagogue-based women's group that made caps and mittens to donate to the school. When her only child was near death due to rheumatic fever, Katz had promised God she would devote herself to good works for sick children if the boy lived. Her son Michael survived, and when she learned about Fernald children who never received visitors, she felt she had found a way to fulfill her pledge.

"I may have been a little naïve. I know the attendants thought I was," recalled Katz decades later. "I said that I wanted to visit a child that nobody comes to see. I don't care who it is. They chuckled and then said they would help me do it."

It took several weeks, but finally on a spring Sunday in 1950, Fernald staff informed Bea Katz they had a child for her to visit. She was told to come to the school on Company Sunday and wait in the day room of the Wallace Building, where physically disabled residents lived. She arrived early, and had to wait thirty minutes while an attendant went searching for the designated child. She returned with a ten-year-old boy with metal crutches strapped to his arms, whose head jerked from side to side and whose legs were so bent and splayed that he could barely walk. A folded piece of paper hung out of his shirt pocket. His hair was matted, and his pants were wet. As he drew close, Katz could smell urine and could hear the sucking sound the boy made as he labored to keep saliva from pouring out of his mouth.

"Mrs. Katz, this is Louis," the attendant announced. She told her that Louis was "one of the worst in the school, a low-grade imbecile with cerebral palsy." His mother had died in childbirth and his father had committed him to the institution. He had never before received a visitor.

She rose, walked straight to where Louis stood, hugged him, brushed his hair with her hand, and kissed his face.

In that first encounter, Katz was confronted by a child who could not answer her questions about his age, his family, or anything else. He couldn't play a game, draw a picture, or tell her if he wanted to go outside. But the day room had a radio, and when Katz brought Louis near to it and tuned it to music, she saw that he liked it. "I saw there was somebody inside there," she said years later.

When the visit was over, Bea Katz went to the Administration Building to find out more about Louis. They handed her his file, and she learned that his last name was Frankowski and that he had been born one year before her own son Michael. He was deemed untrainable and received only custodial care. She also learned that the paper Louis kept in his pocket at all times was a blank sheet. He insisted that it was a letter from his father.

The way Bea Katz figured it, the attendants had expected her to be horrified by Louis Frankowski's appearance and were sure that after that first encounter they would never see her again. They were wrong. Katz came every Company Sunday. She spent hours talking to Louis—telling stories about her son, her husband, and her own life—

and eventually realized that he understood almost everything she said. Sometimes she discovered evidence of what life was really like in the wards. One Sunday, she forgot her purse, and when she returned for it, she found two attendants eating the plate of cookies she had left for Louis. One winter, after she had given the school piles of donated mittens and hats, she learned that instead of distributing them the attendants had taken them outside the institution and sold them.

As she came and went, Bea Katz encountered dozens of children on the grounds. They ran to her, grabbed her hands, and asked questions. Some asked for her help, pleading for her to "call my parents, tell them I'll be good and get me out of here." One girl told her that her mother and father had sent her to Fernald because they were afraid that potential suitors would reject her older sisters if they saw that a disabled child lived in the family.

Freddie Boyce was among the children who approached Bea Katz on their own, and she liked him immediately. He was talkative, energetic, and sincere. "He knew the score," she recalled. "And he was so bright and friendly you couldn't help but fall in love with him."[14]

When spring turned to summer, many of the Home Boys left Fernald for a few weeks with their families. Life in the wards of the BD became a little more relaxed. School was not in session, and the shops were closed. But there was work to do. Each morning the State Boys were gathered up by the gardeners who tended the many acres of corn and the vegetables grown on the west side of the sprawling property. After a long walk to the fields, they were set to work pulling weeds and picking the beans, cucumbers, tomatoes, and other crops. The most intelligent State Boys, including Freddie and his friends, were favored for these chores because they readily recognized the difference between the weeds and the vegetable plants. They also worked faster, and the gardeners always seemed to be on a tight schedule.

The pressure on the farm managers was reflected in a report from this era. "The one difficulty is that this work is not completed when school classes are to begin," noted a farm supervisor. Administrators then allowed him to hold the best vegetable pickers out of their classes for an extra month, so they could finish the harvest.

The loamy smell, the warmth of the sun, and the feeling of the tilled earth beneath their knees would stay with the boys for the rest of their lives. But they would also remember the sunburns, sore backs, hunger, and thirst. During his first few summers in the fields, Freddie Boyce struggled to stay focused on the task at hand, whether it was yanking weeds or plucking beans from a vine. His mind wandered to elaborate fantasies about elves and animals that lived in the garden. Sometimes he would get so lost in these thoughts that he stopped working. Then he would be shocked back to reality by a shout from one of the gardeners. "Wake up and get to work, kid."

At about 4:30 each afternoon, the boys were trooped back to the dorm for a dinner break, but their work was not necessarily finished. Sometimes after their meal, they were taken to the Fernald cannery to clean and sort the vegetables they had picked during the day. The boys sat on crates and snapped the ends off of string beans, pulled stems off tomatoes, and shelled peas. When they were finished, the bushels of vegetables were taken to another room, where female inmates would wash, cook, and then can them.

The combined production of the farm at Fernald and the one at the distant Templeton Colony was carefully recorded. Each year, the State Boys and their supervisors produced hundreds of thousands of pounds of fruits, vegetables, berries, and grains. In a single summer, the tomato crop alone could exceed 70,000 pounds. At Templeton, where the beef cattle were raised, about 1,000 animals were slaughtered each year. The meat was eaten by state wards and state employees. The hides were sold to tanneries.[15]

The hard work done by the State Boys impressed the hired farmhands and foremen. They were not experts in IQ tests or child development and didn't have access to the boys' files, but they could see that many of them were able to follow instructions and complete assigned tasks quickly. As a consequence, the farm operators were far more relaxed than the ward attendants, and so it was easy for Freddie to slip away one Company Sunday and find the pleasant-looking lady he had spied on at the North Building. He caught up to Bea Katz as she pushed Louis in a wheelchair on one of the campus walkways.

"I can push for you."

The voice was a child's, but the tone was assertive. Mrs. Katz turned away from Louis to see a brown-skinned boy with big chestnut-colored eyes, curly dark hair, and dirt under his nails.

"He smiled at me and just started to help push the wheelchair," she recalled. "He was just a regular boy as far as I could tell, mischievous, smart, and my first thought was, Why is he here?"

On that Company Sunday, Bea Katz learned Freddie's full name and listened to him talk about his life at Fernald, about his friends in the ward and about the work in the fields. He was extremely talkative, which she found amusing. He was also completely at ease with Louis.

After that first walk together, Bea Katz decided to include Freddie in her Company Sunday visits. Usually, they did nothing more than walk together, with Louis, but for Freddie these strolls meant that someone on the outside knew he existed and cared about him. Other boys saw this and would sometimes tag along when he went out to greet Mrs. Katz. No doubt one or two of these boys were present when she got permission from the administration for a drive in the car to get some ice cream.

It was 1951, and Freddie hadn't been in a car in more than two years. As he settled into the back seat, he felt excited, and a little afraid. Soon after she began driving, Mrs. Katz noticed puffs of smoke coming from under the hood of the car. She began to pull the car over to the side of the road. Freddie noticed that Mrs. Katz was struggling to keep calm. He also spotted the smoke. Remembering fire drills at Fernald, he acted immediately, opening the door while the car was still rolling. Seeing this in the rearview mirror, Mrs. Katz shouted at him to stop. He burst into tears. It would take fifteen minutes to confirm that the car was all right and to calm Freddie down.

After that incident, Bea Katz became concerned about taking on too much. At least a thousand desperately lonely boys lived at Fernald, and she couldn't soothe them all. Louis remained her primary focus, and she would continue to see Freddie on Company Sundays, but she would not encourage other boys to join them. Over time, she would befriend a third boy, one whose unusual appearance and remarkable intelligence made him unique at Fernald. Named Albert, he was a tiny young man, under five feet tall, whose bodily malformations included a misshapen chest that pointed outward, like the prow of a ship. Like Freddie, he introduced himself to Katz and tagged along on a walk, chatting incessantly and charming her into adding him to her entourage. For years to come, she would watch all three boys mature, noticing progress that the attendants, who came and went, could never recognize. She would tell them that they were smart. This had an effect on Freddie. Slowly the ether was being dispelled, and he began to imagine a life beyond Fernald.[16]

While Bea Katz encouraged Freddie Boyce to reject the notion that he was a moron, specialists in education were in the middle of a national debate over the true significance of IQ scores and the nature of basic intelligence. One stimulus for this discussion was a paper that had been published in 1946 by psychologist Bernadine Schmidt. In 144 pages, Schmidt documented how she had used intensive education over the course of three years to raise the IQs of 252 supposedly "feebleminded" adolescents from an average of 52 to 89, which was considered low normal.

Schmidt's experiment began after administrators in the Chicago school system had become alarmed by the poor performance of students in classes that served slow learners. She had recruited superior teachers, trained them in managing difficult children, and then given them the freedom to teach as they saw fit. The new teachers replaced basket-weaving and other "handiwork" with academic lessons presented in innovative ways. The results were startling. Besides raising the mean IQ of the group, the program actually pushed more than a quarter of the children to scores of 100 and beyond. The improvement continued after they left the program. Half the children attended high school, and half of those earned diplomas. In contrast, the median IQ score of a control group that did not attend special classes actually declined.

The results of the Chicago experiment take on a certain poignancy when one considers the several pairs of twins who were separated, with one getting the special classes while the other did not. In each instance, the enrolled twin thrived in comparison. Typical were Jessie and Jaqueline. Jessie lost six IQ points during the three years she was in the control group and became a truant from the school where her lessons focused on weaving and stitching. As soon as she was old enough, she dropped out of school entirely. In the same time period, Jaqueline's IQ increased from 42 to 89. She became an avid reader and, after the study, continued her education into high school.

"A majority of children classified as feebleminded can grow to be mentally competent, well-adjusted members of a democratic society," concluded Schmidt. She didn't speculate on what her work revealed about the larger issues surrounding such children. However, anyone reading her paper should wonder if the expert assessments of these children, and the labels they received, had been correct in the first place.[17]

Bernadine Schmidt was an educational psychologist, not a clinical psychiatrist, and as such did not begin with the bias of genetic orthodoxy. Her paper shook the psychological establishment, which responded vigorously. The first attacks were lodged against her data, which were found to contain minor flaws. Then other researchers conducted smaller, short-term studies that failed to replicate Schmidt's result. But none of the critiques erased the most important point suggested by Schmidt's work—that seemingly deficient children needed education as much as asylum.[18]

While Schmidt and a few other critics challenged experts in institutions, a confessional article published in 1950 by Nobel Prize–winner Pearl S. Buck made many more Americans reconsider the function of state institutions. First appearing in *Ladies' Home Journal,* Buck's piece revealed her daughter Carol's life at the New Jersey Training School in Vineland, New Jersey. She contrasted what she thought was a positive atmosphere at Vineland with the hopelessness she saw in typical institutions filled with children "sitting dully on benches, waiting, waiting."

Later printed in book form with the title *The Child Who Never Grew,* Buck's piece marked the first time she had even acknowledged Carol's existence. It made a forceful argument that her daughter benefited from the Vineland school's care, and this was a balm to the guilt felt by other parents who had made similar choices. Thousands of them sent her letters of thanks.

A number of later articles, including one by writer Judith Crist, also affirmed the benefits of separating feebleminded children from their families. Some even said that the institutions were the brave, best choice for many children. But inevitably a backlash developed. It came from an unlikely source, the entertainer Dale Evans.

Published one year after Buck's work, Dale Evans's memoir, *Angel Unaware*, tells the story of her severely disabled daughter's brief life. In the book, Evans explains why she kept the child at home and insists that her daughter, who died at age two, had brought her family more joy than pain. It was the third-best-selling book of 1953.

No matter where they came down on the subject of institutions, the books and articles made the lives of retarded children better known and removed some of the stigma attached to their families. Parents whose children lived in state schools began to organize for mutual support and to pressure government agencies for services. In New York, families of retarded children formed an organization that eventually became the Association for the Help of Retarded Children. Clemens Benda was an early adviser to this group. He encouraged them to press politicians for services that would extend care and training over a child's entire life. In 1952, parents in Massachusetts founded the Fernald League to push for better care for retarded residents of the school. Two years later, responding to individuals like Bea Katz and social service groups, Fernald established a department of volunteer services, which brought in even more outsiders.[19]

Unnoticed in all the publicity and community response to the problems of institutions were the thousands of *normal* children who were being diagnosed improperly and then locked away by officials who could find no other place for them. These lost boys and girls, including Freddie Boyce and his peers at Fernald, were easily overlooked when activists considered the bent and twisted bodies in cold, overcrowded rooms, or children with Down syndrome denied a chance to reach their potential. Boys with bad parents, and delinquents who were slow learners, just didn't evoke the same kind of sympathy.

Paradoxically, as more outsiders became interested in the Fernald School, the rising population in the institution was making conditions in the wards—the places outsiders never saw—much worse. In the BD, as many as fifty boys were being jammed into wards that were built for thirty-six. The staff was not increased to accommodate them, and attendants found it ever more difficult to maintain order. This was especially true when it came to Freddie's group of friends, some of whom had gotten old enough to enter puberty. So much energy boiled inside them that they just couldn't spend long hours on the settees with their arms folded. Silly jokes would fly around the room, or one boy might wiggle against another. A shove would be followed by some protest. Soon a wrestling match ensued. All this was normal for boys but unacceptable at Fernald.

When he was in charge, attendant McGinn was quick to order an entire ward of boys to kneel on the iron frames of the beds for an hour or more. When they came off the iron, the boys couldn't walk. This assured peace and quiet at least until the effect wore off.

Years later, former State Boy Larry Nutt described another form of punishment McGinn employed to cause pain without leaving marks. He made the boys line up like a company of soldiers and then told them to drop to their knees. Each boy was then required to place his hands behind his back. Still kneeling, the boys were then told to lower themselves backward, until they were lying on the floor, hands still behind them, knees still bent. The knee strain caused severe pain and then numbness.

Unlike the bed-iron treatment, which required the boys to keep their balance, this knee-bending could be maintained for hours, keeping the boys immobile. This may have been why McGinn favored it. He would get the boys into this formation and then walk up and down the rows lecturing them on obedience and proper conduct. If he had another ward to mind, he could leave the boys in position, check out the other room, and return to find things as he left them.

In this state, the boys lay staring up at the ceiling and the rows of lights. When McGinn thought they had had enough, he pulled a straight pin from his pocket. "I am going to drop this," he announced. "If I hear it hit the floor, you can get up. If I don't, then you stay where you are."

*   *   *

Sometimes a State Boy who lived with abuse at the BD might find sympathy elsewhere at Fernald. At a new hospital that was constructed near the Administration Building, nurses seemed delighted to care for the boys who came over from the BD. They were engaging and bright, especially when compared with the seriously disabled infants and children—many of whom were slowly dying—who occupied most of the hospital.

Robert Gagne, who spent more than a week in the hospital with German measles, would remember for the rest of his life the nurse who brought him comic books and little boxes of maple sugar candy. The stories he told the others upon his return made them all pray to develop the red spots and fever that might bring them to the hospital.

A fever wasn't required for the boys to receive attention from schoolteacher Kenneth Bilodeau. In 1950, Bilodeau was a twenty-four-year-old recent college graduate who had absorbed a more modern approach to psychology during studies at Suffolk College (later called Suffolk University) and Boston University. Though he started at Fernald as a psychologist, he soon transferred to fill a teacher vacancy at the school. He was assigned a clutch of higher-functioning boys— Freddie, the Gagnes, Joey, and others—who had finished sense training and were deemed ready for regular classroom work. He was immediately impressed by their abilities and the circumstances that had brought them to the institution.

"These were kids who never had a constant adult presence in their lives. They had fallen three or four years behind in school and boom! They were sent away," he would recall.

Determined to become the consistent adult presence the boys had missed, Bilodeau abandoned the idea of a seven-hour day, or relationships limited to schoolwork. When one boy refused to get a haircut because he was afraid of the staff barber, Bilodeau did it. When others said they had never ice-skated, he arranged for the school to acquire secondhand skates, dammed up a small brook on the school's property, and then taught the boys how to skate on the frozen pond.

Boys like Freddie, who were hungry for attention, responded immediately to Bilodeau and became obedient, even pliant. Each one of them was desperate to feel recognized as an individual, to be truly known and appreciated by a kind adult, and Bilodeau represented their best chance. Freddie competed for his attention by trying to answer every question in class. Others tried to impress the teacher in the gym or on the playing field.

Bilodeau employed more direct methods to reach those boys who were less self-controlled and perhaps more hardened by their time in the institution. Two brothers posed a problem typical for this group. They both liked to sneak up behind a teacher or attendant and then embrace him with a bear hug so tight that it would be difficult to breathe. A young military veteran, Bilodeau was prepared the day one of these boys grabbed him from behind on the baseball field. He jerked himself downward, reached between his own legs to grab the boy's ankles, and pulled so hard that the kid tumbled onto his rear.

A quick wrestling trick could break one boy's bad habit, but Bilodeau would have to find a more reliable way to let the growing boys blow off steam. With the help of the gym staff, he let boys challenge him to boxing matches. One of the first was perhaps the toughest in the group, Jimmy Hannaford. Two pairs of gloves were available, a large set and a small one. The boy chose the bigger gloves, not realizing that the padding would cushion his punches. After he felt the sting of Bilodeau's small glove, and reported the pain to the other boys, the challenges stopped. But the boxing ritual remained available for boys to settle problems among themselves, and they often used it.

On one of the few occasions when Freddie and Joey had a falling out, Bilodeau put them in the ring to resolve it. Freddie had grown taller and a bit stronger, and Joey was slow to lace up the gloves because he knew what was coming. Though skinny and generally easygoing, Freddie had a strong will and hated to lose at anything. Joey put up his hands, tapped Freddie a couple of times, and then took a right hook to the head that landed him on his back. Shocked by what he had done, Freddie quickly knelt on the floor to tell his friend he was sorry. They never had another serious disagreement.[20]

Although he considered their emotional needs, Bilodeau's main task was to move the State Boys along academically. The school at Fernald did not measure up to public school standards. Books—ancient castoffs from other schools—were in short supply. The teachers were not required to have even the most basic credentials, and many did not. Many of the students suffered from what later would be called learning disabilities or attention deficit disorder. Those who didn't have these kinds of problems had fallen so far behind their age group that Bilodeau had to spend long months on the basics, including the alphabet, learning to use crayons, learning colors and shapes. The boys knew this was "baby work" and often resented it.

Freddie was still doing work at first- and second-grade levels when he was eleven years old. The same was true for many others. Most had never learned to read or to perform simple arithmetic. But all the boys were eager to learn how to look and act more grown up, more like boys on the outside. Often this interest far outweighed their academic curiosity. Bilodeau first noticed this at the start of an otherwise unremarkable day. He was giving instructions about some activity when he saw something strange about the way his boys were dressed. Each one had shifted his belt so that the buckle was on his left hip. The teacher was puzzled until he recalled that for gym classes he put his own buckle in the same spot to protect boys who might stumble into him during activities. The boys were making him their role model, in hopes that they might know how to behave—right down to the proper spot for a belt buckle—should they grow into men with lives on the outside.

More could have been done for the brighter residents of Fernald if they had full-length school days. But they spent less than half as much time in the classroom as public school children, because their days were split between the school and various shops. During this training, called "manual" for short, the boys made items that the institution needed to function and learned skills that supposedly prepared them to work in industrial shops in the years to come.

Manual training began with elementary lessons. A state school paper titled "Industrial Training for Imbecile and Moron Boys of School Age" describes how those at a three-year-old level would start by chasing balls scattered by a teacher. They would move on to making piles of sticks and stones in the schoolyard, stringing beads, and placing pegs in pegboards. By the time they reached the "five-year mental level," the boys in manual training would be making brushes and working on so-called "Todd Looms."

The Todd Looms at Fernald were five-foot-tall wooden frames with scores of strings running from top to bottom. Boys sat on little wooden stools and wove colored cloth strips in and out of the strings to make a rug. Some children got so good at this work that they would advance to large, industrial looms. There they balanced on an elevated bench and used a shuttle, pedals, and a big lever to draw thread from huge spools to make all different kinds of cloth.

In the brush-and-broom shop, children sized and cut straw, fastened it to wood forms, and then added handles of various lengths. In the print shop, they produced everything from stationery for the state legislature to traffic tickets for local police departments. In yet another manual room, the boys assembled the wooden hangers that some attendants in the wards used to give them beatings.

The finished goods made by the children, including resoled shoes, stationery, textiles, furniture, brooms, and brushes, were used at the school or sent free of charge to state agencies such as the courts and the department of motor vehicles. The output was substantial. In a given month in the 1940s and 1950s, Fernald might produce 350 bed sheets and towels, 150 brushes, 25,000 sheets of stationery, and scores of wooden objects.[21]

After one full year with Mr. Bilodeau, Freddie Boyce had become a whiz at manual, weaving tight braids for rugs and quickly installing bristles in new brushes and brooms. He was also beginning to talk more, opening up emotionally. Mildred Brazier, the school's longtime principal, wrote that he was "cooperative and good natured." She also noted that he talked so much "that it has created a problem in discipline." An IQ of 79 was posted above Brazier's notes. This represented a jump of nearly 10 percent over his IQ at admission, eighteen months prior.

Although Miss Brazier seemed unable to see Freddie's progress for what it was, an indication that he was not a moron in the first place, it did not go entirely unnoticed. At about the same time, a psychologist would write her own brief description of Freddie that proved that someone noticed his progress: ". . . the quality of his work is excellent on all types of questions calling for independent reasoning," she wrote. "He initiates conversations on matters pertaining to his surroundings and on brief acquaintance seems a well-adjusted little boy."[22]

*   *   *

Freddie Boyce was learning to present the world with his version of a normal boy, which was based on lessons from Mr. Bilodeau, glimpses of television, the behavior of his mates in the BD, and his own experiences with adults. Freddie understood that he was on his own, and that Fernald was his world. For this reason, he was better able to focus on the task of building an appropriate image than were those boys who worried constantly about their families and obsessed over who might visit. They had known mothers and fathers and homes of some sort, and continued to hope that they might be rescued. Joey Almeida pined for contact with his family, and so did Robert and Albert Gagne. Their longing was so strong that it even weighed on the hearts of some of the attendants, who noted in official records that as the months passed no one ever came to visit Robert and Albert. Then, in the summer of 1951, they were surprised to hear that Ruth Gagne had written to ask about visiting hours.

Two and a half years—thirty Company Sundays—had passed since the Gagne boys had been deposited in the BD. The letter was the first sign that anyone on the outside was interested in their fate. Attendants mentioned the contact to the boys, and they became even more desperate for a visit. But one month, then another, and another came and went with no one calling for them. They wouldn't learn until they were adults that on one of those Sundays their mother had attempted to see them, but she appeared long after visiting hours were over and was sent away.

Finally, on a Sunday in October, Thomas and Ruth Gagne managed to travel from Newburyport to Waltham and reached the Boys Dormitory by early afternoon. They checked at the office on the first floor of the dorm and were directed to the day room. Their sons were summoned from the ward upstairs. Others who were there would recall that one of the Gagne boys had said, "I told you she would come." Together, Albert and Robert raced ahead of the attendant and ran down the stairs.

In the day room, Mr. and Mrs. Gagne braced themselves to see their sons again. But when the boys came to the door, they did not recognize their parents. Ruth had to call out to them, and even then they were confused. The thrill they had felt coming down the stairway quickly changed to anxiety and fear. Slowly, they came to her, and she hugged them. Their father gave them the comic books and candy he had held in his lap while waiting. They spoke for a few minutes, went for a walk around the grounds, and then were gone.

After that first brief visit, Robert Gagne began to feel not closer to his parents, but more distant. The fantasy of a mother and father who loved him and would come make things better had been replaced by the realization that he hadn't even recognized them. The visit was brief, uncomfortable, and ended with no assurance that his parents would ever come back.

Thomas Gagne would try to visit one more time but would be turned away because he arrived drunk and was carrying a bottle of whiskey. Ruth Gagne would return a few times. Albert, who had spent more time with her before coming to Fernald, welcomed her. But Robert lost interest in the visits. (He understood that his parents had been responsible for his confinement at Fernald and believed they would do nothing to get him out. For these reasons, he chose not to see her.)[23]

Further evidence of the trouble that besieged the Gagne family arrived at Fernald in the fall of 1951. On a Sunday, a clerk in the Administration Building summoned Robert and Albert to Dr. Farrell's office. The superintendent brought them into his conference room, where they saw a slim, dark-haired girl of sixteen. She was their older sister Doris, who had just been admitted.

The boys rushed to their sister. She embraced them and then started to cry. In the six years they had been apart, she had almost forgotten them, but with the reunion, the pain of the separation and of memories of life in Newburyport rushed back. She was surprised to see how Robert had grown and shocked to see that Albert hadn't. Though thirteen years old, he was just four feet seven inches tall and weighed only seventy-five pounds. He could have passed for a normal eight-year-old. They had only a few minutes together to catch up on everything that had happened in their years apart.

Doris told them that after social workers had emptied the Gagne home of children, she had remained in foster care but never attended school regularly. She had been brought to Fernald by a social worker following a confrontation with a foster mother. The woman, who Doris said was prone to slapping, had beaten her, demanding that she admit to something she didn't do. Finally, Doris fought back, pushing the woman to the floor. The foster mother called the police, who arrived with a state social worker. They brought her to Fernald, where she was taken to a testing room for an evaluation.

"This woman, Miss Chipman [psychologist Catherine Chipman] came in with this cane, dragging her leg," recalled Doris in her adult years. "She gave me these tests to do, and I just wasn't in my right mind. It was working with blocks and folding paper and other things. She scared me, and I was shaking. I guess I failed, because the next thing I knew, they were taking me over to the Girls Home and putting me in a ward."

At the GH, Doris discovered that most of the girls were, like her, ordinary teenagers from troubled backgrounds. A particularly tough girl named Betty ruled the ward where Doris was given a bed, and she immediately tried to bully the newcomer. When the two were sent to fold blouses in the clothes room, Betty kept knocking over the piles that Doris completed.

"You want to hit me, don't you?" she hissed. When this didn't work, she came around the table where they were working and shoved her.

"C'mon, do it."

Having already recognized that attendants were the real power in the ward, Doris resisted the urge to fight until one of them appeared and ordered Betty back to work. In the months to come, Doris carefully avoided being alone with Betty, and the bully realized that she couldn't trick Doris into getting into trouble. Like her brothers over in the BH, Doris was able to gradually demonstrate to the attendants that she was one of the "good ones" who wouldn't cause trouble.

During their first meeting at Fernald, Doris had told her brothers that she planned to escape one day and that once she was free she would then do what she could to get them out, too. In the months that followed, Albert and Robert sometimes saw their sister as they walked the grounds. They were permitted to sit with her, on the girls' side of the Howe Hall

Auditorium, during a Saturday movie. And once, when the music department of the school put on a Christmas pageant, Doris, who had a beautiful voice, sang "O Holy Night" with Albert on her lap.

A model citizen of the Fernald School, Doris took direction so well that she was assigned to work as a maid at the superintendent's house. To her, the large, well-furnished house seemed like a mansion, and the Farrells, a remote, privileged family. She helped in the kitchen and cleaned the rest of the house. When she was seventeen, she was paroled to work as a live-in maid for a family that lived off the grounds.

In the first months of her parole, Doris made sure to visit her brothers on Company Sundays, taking a bus to Waverley Square and then walking up to their dorm. Between visits, she began to chafe at the household rules, which required her to work long hours and barred her from dating or going anywhere but to see her brothers. Finally, on the first Sunday in July 1952, she went to the Boys Dorm and talked the attendant into letting her see Albert and Robert, even though it wasn't an official visiting day. The boys were brought downstairs, and the three sat on a bench outside the office. It was the same bench where new boys waited while their admission forms were being filled out.

"My sister had everything timed out, so she didn't have long to talk to us," recalled Albert Gagne. "But she said she was leaving that day. She also said that she couldn't promise anything for sure, but that she would try to help us get out. She stood up to hug us both, and then she was gone."

Doris went back to the home where she worked and changed into a pair of shorts and a short-sleeve top. Late in the afternoon, she slipped out of the house and walked down to the road. When a car approached, she stuck out her thumb. The young man driving the car stopped and asked where she was going. She asked where he was headed. "Buffalo" was the answer. She said that sounded good to her and got in.

While her brothers finished another day at Fernald, ate dinner, and prepared for bed, Doris listened to the car radio and gabbed with the man who had picked her up. Traveling along Interstate 90, they passed Worcester, Springfield, and then Albany, New York. He bought her dinner, and when he was tired, he parked to rest. It was then that he made a clumsy pass at her. She fended him off. Embarrassed, he told her that she didn't have to worry. He understood the word no.

The next day, when her brothers were in the schoolhouse at the institution, the young man who had stopped his car for Doris back in Massachusetts drove into downtown Buffalo. He explained that he was going to Canada and couldn't take her along. As he approached the Peace Bridge, which crosses the Niagara River and connects Buffalo to Fort Erie, Ontario, he pulled over to the side of the road. Doris got out and stood on the sidewalk and watched as the car drove away. She had no idea what she would do with herself, how she might find food and shelter for the night. But she wasn't afraid. She felt free.[24]

FOUR

By the time they were ten or eleven, the State Boys understood that nearly everyone on the outside considered them to be "retards." This word hurt them as much as the word "nigger" hurt blacks. When they were angry, they flung it at each other.

Attendant McGinn reduced boys to tears by calling them retards as they waxed the floors and buffed them with the rope-rubber. More than one would recall, as adults, how McGinn whispered into their ears that they were "worthless" or "stupid" and that "no one gives a shit about you."

Another bit of torture, which McGinn began to use after Freddie had been at the BD for a couple of years, was reserved for those who talked during meals in the downstairs dining room. He would grab a slice of bread from the boy's tray and tear off enough to wad into a ball the size of a large marble. He would then yank the boy to his feet, and order him to get down on the floor and push the bread with his nose. McGinn would laugh and say, "Look at the retard."

Though a few boys laughed nervously, those who watched the spectacle were torn by different emotions. Some felt relief knowing that someone else was McGinn's target. Others felt pity for the boy who was being humiliated. But nearly all felt anger. They were angered by the abuse, which had become commonplace, and by the feeling of dread it produced in almost every waking moment. Day by day, life at Fernald was feeding the rage inside of them.

Though he could get away with the bread trick in front of other attendants, McGinn saved his most humiliating discipline technique for times when he was alone with the boys. He would pick out someone who had been talking, or laughing, and make him strip off all his clothes and stand naked on a table. This caused an uproar among the other boys. Some laughed. Others, especially the Big Shots, joined in McGinn's taunts. Albert Gagne was once a victim of this treatment. He recalled years later how McGinn made the other boys jeer louder by tying a ribbon on his penis. "I think the idea was to humiliate me and use my friends to do it," said Albert. "I don't know if it's possible to feel more ashamed."

The humiliation and constant name-calling—retard, lifer, moron—were difficult to ignore. This barrage beat down Albert Gagne until he began to believe that he *was* defective and destined to spend his entire life inside the institution. He became more and more withdrawn from the other boys, until the summer when he and his brother were given the job of delivering mail to staff throughout the institution.

"They gave us hundreds of letters, and we were supposed to get them delivered by a certain time," recalled Albert years later. As he spoke, he compulsively smoothed the cloth that covered the table inside the kitchen of his tidy house in rural Maine. "Robert and I really worked. We put the letters in order, by building and then by floor, and then went from place to place in a very logical way. Then all of a sudden, all the letters were gone. My heart was filled with joy, it really was, because I realized we had done it, and if we did that, we couldn't be stupid. We had to be smart, to do all that."

The jobs that the Fernald staff gave the boys made them more confident in their abilities. Freddie assisted the janitors who cleaned the schoolhouse every day, and learned that he could do every job they did, from sweeping the floors to mopping bathrooms. The only difference was that he was more thorough. Joey had one of the best jobs on campus, as a baker's helper. He reported early each morning to the institutional bakery and spent three or four hours lugging sacks of sugar and flour and then cleaning up after the adult staff. Unlike the attendants who managed the wards, and the professionals who tested and diagnosed him, the bakers treated him like a normal boy, which only made him more certain that that was what he was. Other boys had similar experiences at their jobs. In general, the service staff was more willing to regard them as normal, perhaps because they saw their strengths, or because they were less concerned with IQ scores and academic skills.

Albert and Robert Gagne were too timid to voice what they believed about themselves—that they didn't belong in Fernald—to anyone on the staff. But Freddie Boyce often demanded to know why he was in Fernald when, as anyone could see, "there ain't nothin' wrong with me." For the most part, attendants ignored his complaints the way prison guards ignore inmates when they protest their innocence. They wouldn't even consider that the doctors who had signed the papers that authorized a boy to be locked in Fernald might be wrong.

Freddie made a pest of himself in Kenneth Bilodeau's class, hoping to prove he wasn't retarded by asking the toughest questions he could imagine. This led to long discussions on the mystery of infinity, the existence of God, and the origins of the universe.

The questions that Fred posed, and his insistent attitude, pleased Bilodeau because they showed that he was becoming more comfortable in conversations. Though the interruptions were often ill timed, Bilodeau was always able to answer quickly and return to his lesson plan. Fred finally stumped him with a question about the bird that held the record for long-distance flight. Freddie refused to accept that his choice—the arctic tern—was wrong. Bilodeau agreed to look up the answer, and the class cheered when he read from an encyclopedia that Fred was right.

Though Bilodeau treated him respectfully, no amount of classroom attention satisfied Freddie's desire to have his intelligence recognized and affirmed. Frustrated, he began to rebel in small ways. In the dining room, McGinn once picked a piece of bread off his plate and began to squeeze it into a ball. All the boys in the room stopped eating and focused on the little drama about to be played. McGinn put the wad on the floor between two tables and then turned to Fred.

"Get down and push it."

Fred got up from his chair and kneeled on the floor. But then he stopped, frozen between his fear of McGinn and his pride.

"C'mon, little nigger boy."

"No, I won't," answered Freddie. He looked at McGinn and tried to appear calm and determined, though he was shaking.

Standing over the boy, McGinn jingled his keys, signaling that he might whip Freddie with them, but instead he ordered Freddie to stay there, kneeling, until the meal was finished and the group was dismissed. Some left believing that McGinn was going to punish Freddie in private. He did not. Instead, he then told Freddie to leave, but keep his mouth shut. He never tried to punish him that way again.

Signs of Freddie's rebelliousness appeared more and more often. In the manual training department, where he was supposed to learn how to paint furniture, he became disinterested and sloppy. Mildred Brazier, the school principal who a year earlier wrote that he was "cooperative and good natured," now reported that he was "sneaky" and "defiant."[1]

As they were growing older, many boys began to show their anger. They would continue talking when told to be quiet, or stall when instructed to do some chore. School officials became so alarmed about discipline in the classrooms that they began reporting the worsening mood of the school to the Fernald trustees. Escapes rose, and so did the number of Fernald residents placed in restraints and in isolation in Ward 22.

Restraint and isolation reports, issued monthly, were an accurate measure of trouble among Fernald's residents. Throughout the 1940s, these statements showed an average of three or four residents in seclusion and none in restraints. In 1950, the numbers began to rise, and it became common for between a dozen and twenty to be locked up in a given month. By 1952, the daily population of the prison ward averaged roughly two dozen, and on any given day six or more Fernald students were bound in leather straps or straightjackets.[2]

The increase in severe punishments could have been blamed in part on the rising population of the institution. Fernald's enrollment reached 2,032 in 1952, and would increase to 2,242 in 1954. The place was overflowing with people. But the turmoil reflected in the reports was also due to the large number of more intelligent students who were coming of age in the wards and questioning why they were being held like prisoners. They knew that their education was abysmal and that they were missing out on the excitement and promise of teenage life on the outside. They feared that if they ever did get out of Fernald—there was no guarantee this might happen—they would be old men unable to cope with normal life.[3]

Some of the State Boys went beyond questions and prepared themselves to challenge attendants physically. Eric Johnson* was the first to do this. By the time he was thirteen years old, Johnson was sneaking into the basement of the dorm and doing countless push-ups and sit-ups. Eventually, he fashioned barbells out of a pair of milk cans, which he filled with water. He kept this contraption in the basement and went there every day for hours of exercise. He became bigger and stronger every month.

When he was big enough, Johnson began to use his muscles to defend smaller boys from the bullies, intervening when they picked on Albert Gagne or one of the Dopes. Then a new attendant, a former U.S. Marine, was assigned to the ward. He immediately identified Johnson as a competitor, or threat, and tried to intimidate him. One morning, when the attendant found that a quarter wouldn't bounce off of Eric's bed, he went around the ward yanking all the blankets off the beds. He ordered Eric to make up every one of them and then went out the door.

With all the other boys watching, Eric ran to the stairway. The attendant had just reached the first landing on his way downstairs. Eric flung himself down, onto the former marine's back, and the two collapsed onto the landing. The marine's face was bloodied before another attendant arrived and broke up the fight. Afterward, Eric became the first of Freddie's friends to go to Ward 22. Always tight-lipped, he said very little about the experience when he returned to the BD with his head shaved. He did say that he had not received the dreaded shock treatments or a lobotomy. In fact, Ward 22 wasn't nearly as bad as he expected it to be. The other boys considered him their hero.

Although incidents like Eric Johnson's attack on the marine survive only in the memories of those who were there, records at the Fernald State School offer evidence of increasing problems on the campus. According to Farrell's reports to the trustees, qualified workers were becoming so scarce that when an attendant was sick, replacements were often unavailable. In one instance, a male attendant was allowed to keep his job even after it was discovered that he had just been released from prison, where he had served time for violating the Mann Act, which prohibited transporting persons across state lines for criminal purposes.

In another incident, in May of 1952 a physician making her rounds discovered that a nine-year-old resident named Joseph Tersigni was covered with welts. State Police were called to investigate and they concluded that an attendant named Trudy Blacksmith had beaten him. Farrell fired her.

The disarray in the school was apparent to a couple of trustees who made surprise inspections and found the wards and dining halls to be dirty, overcrowded, and understaffed. In the summer of 1952, flies infested the school infirmary. Staffing was so inadequate that a severely retarded child was left unsupervised long enough to die from swallowing a ball. In the overcrowded Girls Dormitory, a single attendant supervised eighty children. In the winter, the heating systems in many buildings were so poor that even with extra blankets, boys and girls shivered through the nights.[4]

In the Boys Dormitory, the beds were moved to within inches of each other. Boys had to walk across adjoining beds to reach their own. Freddie and the others could see that the attendants—who began to shout and threaten them more often—were overwhelmed. To undermine them, the boys pretended that they didn't hear an order, requiring that it be repeated over and over again. They played tricks whenever they could. A favorite involved snatching any keys an attendant might put down for a moment and then hiding them or using them for a game of keep-away.

Every act of defiance was a challenge to the "retard" label. In a few rare cases, these challenges worked, and some on the Fernald staff were forced to recognize that a diagnosis or assessment might be wrong. Soon after he was admitted, Joey Almeida began insisting that the state had gotten his age wrong, that he was really ten, not eleven. In a meeting with a social worker, he was so adamant that the social worker finally agreed to look it up. It turned out that Joey was right.

Intrigued by the way Joey had calmly asserted himself in conversation, the social worker gave him a new IQ test. In his subsequent report, he wrote that he found "no real evidence of this boy being significantly retarded, particularly to a degree that requires institutionalization." Joey's problem "seems to be emotional rather than his being retarded . . . if guided correctly, in a place other than Fernald, he would have a better opportunity in life." Nothing was done in response to this report. Joey remained in Fernald, in the Boys Dormitory, perpetually worried that the daily taunt—"You're a lifer"—was his fate.[5]

The State Boys were so conditioned to believe that they would never be free that it came as a shock when one of them was suddenly released. The episode began with the arrival of a new boy named Bruce Honeysett. Already depressed and angry when he got to Fernald, Honeysett kept to himself. In the ward, he insisted that he didn't belong at Fernald and he avoided making friends. The other boys resented him for being aloof and for thinking he was better than them.

In Kenneth Bilodeau's classroom, Honeysett was able to solve every math problem. He read very well and knew the answers to all the questions the teacher asked, whether it involved the solution to a simple equation or the definition of a word. Recognizing the boy's ability, and his boredom, Bilodeau brought in more advanced books and allowed him to sit and read while the others worked on their lessons. These little interventions did not help much, and Bilodeau worried that the new boy might never adjust. He also worried that he *would* adapt and become mentally deficient due to the Fernald environment.

Honeysett fell deeper into depression. He developed a habit of removing his shoestrings and using them to make little tourniquets around his wrists and fingers, which he tightened until parts of his hand turned red, blue, and then nearly black.

Bilodeau took away the strings before he did any real damage to himself. At the end of the day, the boy got the laces back. The next morning, the whole thing started over again. After about a week, Bilodeau kept Honeysett after class and asked him why he was hurting himself. He said he was bored and it was "something to do." He then told Bilodeau how he had come to be at Fernald. He recalled that he had gone to school one morning, as usual, but returned to find no one home. Afternoon turned to night, and still his parents didn't appear.

The teacher never found out whether the boy's parents had simply abandoned him, or whether they had met some other fate, which had left their son an orphan. Either way, state authorities had taken custody of the boy but, rather than putting him in foster care, determined he belonged at Fernald. Perhaps he had been in shock, or was so bewildered and depressed that he had seemed feebleminded during an evaluation. Whatever the explanation, nothing about Bruce Honeysett suggested that he was retarded. To Bilodeau's eye, he actually appeared to be of above-average intelligence.

This time a teacher's report to the administration produced results. Bruce Honeysett was given an IQ test, and he scored 90, which meant that in ideal conditions he could have scored at least 100. The test so impressed Superintendent Farrell that he approved Honeysett's release from Fernald and transfer to the foster care system. In a matter of days, he was gone. (Unfortunately for the brighter boys he left behind, Farrell was not inspired to look for others like Honeysett.)

Bruce Honeysett came and went so quickly that most of the boys in the BD never spoke to him. It was unlikely that any of them actually missed him in any way. But the fact that he had gone someplace else, presumably someplace better, contributed to the allure of the outside. With each day, the State Boys grew bigger, smarter, and more obsessed with what lay beyond Fernald's boundaries. They talked incessantly about what they would do if they got out. Some also became more determined to do it.[6]

The large ventilator shafts that made the Boys Dormitory look like a brick castle with four turrets on top ran from the basement to the roof. On each ward, the shaft occupied a corner of the room and brought in fresh air, which came through a four-foot-square opening cut at floor level. The opening was covered with a metal screen, kept in place by wing nuts.

No one would remember who first unscrewed the nuts, pried off the heavy wire screen, and peeked inside the ventilator shaft. From the second floor, he could look up to light coming in where a square roof cap, open on all sides, kept rain out of the shaft but let the air flow. If he looked down, he saw more light coming from the basement, where pipes carrying steam heat and hot and cold water entered the shaft and ran up.

Boys had likely explored the ventilator shafts ever since the BD was first occupied, but Freddie's group didn't discover that they could use them as escape routes until the summer of 1952, when Freddie was twelve. They waited until the attendants were gone—usually at night—and then quietly removed the grate. They then climbed inside. To keep from falling straight to the ground, they had to brace their backs and feet on the walls, exert a significant amount of pressure, and find a handhold on the pipes. This last move was tricky because some of the pipes carried scalding steam and hot water.

At first, Freddie just went into the shaft to test his ability to stay safely in place. Gradually, he became bolder. He inched his way down to the first floor, where he listened to people in the day room and struggled to keep from laughing as he spied on them through the screen. One night, he climbed all the way to the top and found he could wriggle out under the roof cap and onto the slate roof of the dorm. He could see that from there it would be possible to climb over the edge and shimmy down a drainpipe to the ground. On another excursion, he went down into the basement and found that he could enter a utility tunnel that ran under the quadrangle that separated the BD, the schoolhouse, and the Girls Dorm.

These adventures made Freddie feel as if he were one of the action heroes he saw on television. Eager to share the fun, he sometimes enlisted Joey as a companion. Fred enjoyed putting together little schemes to visit the GD or roam the grounds at night. But Joey was actually bolder about certain risks. He didn't worry so much about being discovered or punished. Sometimes he even thought it would be a nice change to get sent to the discipline ward. At least it would be a new experience.

Freddie and Joey weren't alone in their daring. Some of the Big Shots tried the ventilator shaft, and some bragged that they had gotten outside and spent time in the nearby town of Waltham. They then crept back into the dorm before anyone ever noticed they were missing.

The boys who sneaked out didn't always get away with it. In November 1952, an attendant armed with a flashlight made an early-morning check and found three beds empty and the grate askew. She moved it aside, put her head in, and shined the light up to see Freddie and two others—all in night shirts and bare feet—working their way toward the roof.

The lights came on in the ward as other attendants came to help and all the boys who had been sleeping awakened. Many of them needed to go to the bathroom immediately, and others began asking questions about what was going on. When Freddie and the others finally came down and through the opening in the shaft, they were met with cheers. Freddie beamed like he had just hit a home run to win a big game.

The next morning, Freddie and his coconspirators were not allowed to dress and were confined to their beds as punishment. This gave him time to think about how he didn't belong in Fernald and to daydream about escaping. These fantasies never involved finding his parents or a permanent home. There was nothing so ambitious in them. Freddie just felt compelled to defy Fernald in any way he could. Running away was the best form of protest he could imagine.[7]

The State Boys' feelings of resentment grew more acute when holidays arrived and most of the Home Boys were retrieved by their families for a few days. Many of these boys were truly retarded children whose parents had placed them in Fernald expecting care similar to that which Pearl Buck described in *The Child Who Never Grew,* her book about her daughter and Vineland. For all they knew, since they were barred from every part of Fernald except the day rooms, the BD was perfectly clean, well equipped, and well run.

For those boys who remained in the institution, Thanksgiving brought a turkey meal complete with sweet potatoes, cranberry sauce, and plum pudding for dessert. Since it was all prepared in a central kitchen in enormous quantities, and continuously heated, the food was almost tasteless. Served in the downstairs dining room, the meal lacked the warm ritual of family and friends. It was more like Thanksgiving dinner straight out of a vending machine.

For Christmas week at Fernald, the school put on a concert and pageant. Boys and girls trooped onto the stage of the auditorium to sing carols. In the days of Walter E. Fernald, these were elaborate events involving a full orchestra, several choirs, a glee club, and a rousing sing-along. By the 1950s, more modest pageants were the norm. Children sang and donned costumes to perform in skits. Then, on the last day before a ten-day school vacation, many classes had small parties in rooms decorated with paper cutouts of Santa Claus and evergreen trees. Some families sent packages of gifts, and the school accepted homemade goodies from women's service organizations, which were given to the State Boys. Each one also received a packet with a few pieces of candy.[8]

With a dozen or so empty beds, the BD felt colder than usual at Christmas, and in some ways the new television made things worse. It brought into the day room scenes of American families feasting or gathered around a tree. It reminded the boys of what they were missing. Nevertheless, they craved time in front of the television and counted the days leading up to the ward's turn to commandeer the set in the room.

On Christmas Eve 1952, the boys on Freddie's ward were due for an after-dinner hour of television. (Attendants had persuaded them to behave by offering them the extra hour if they managed to get through the day without fighting among themselves. Given the emotions of the holiday season, this was hard for them, but they had done it.) But when the shift changed, and new staff came on duty, the daytime attendants failed to mention the deal to the part-time employee who had been drafted into Christmas Eve duty.

After dinner, as they prepared to go to the day room, the boys told the attendant about their extra hour of television. It was their Christmas treat, they explained, a reward for being good. But he hadn't been told of any deal and was irritated about working the holiday watch. His "no" led to begging, and then outrage among the boys. Cries of "They promised" evolved into "You have to give it to us." The attendant didn't have to give them anything, and ended the protests by canceling all television for the night and sending the boys to bed at the regular time.

When the lights went out, many of the boys seethed. As he lay in his bed, Freddie began to think about the day to come and how like all Christmases at Fernald it would include another turkey dinner and gingerbread cake made in large sheet pans. He knew that the cake was already prepared and just waiting in the big central bakery, which was located in the basement of a service building about five hundred feet away. He glanced over at the ventilator shaft.

A single attendant was on duty to cover the entire building. Freddie forced himself to stay awake until, at ten o'clock, the attendant walked through the ward shining his flashlight on the beds and counting the bodies. When the attendant left and walked upstairs to check more wards, Freddie knew he wouldn't be back for three hours or more. He might even fall asleep in the office downstairs. Freddie got out of bed and shook Joey's shoulder. When he was fully awake, Freddie asked him if he had seen the cakes in the bakery when he was at his job. Joey knew exactly where they were and, better yet, he knew about a window with a broken lock where they might get inside.

As Freddie and Joey searched for clothes in the big closet, Eric Johnson awoke and came to see what they were doing. Eric was a little older than most of the boys in the ward and a lot bigger than all of them. When Freddie told him about the plan to steal a cake and bring it back to the ward, he wanted to help. Together they decided that if the night attendant returned before Freddie and Joey, Eric would do whatever he could to keep him from discovering the two were gone.

With the plan set, Freddie removed the screen that covered the ventilation shaft. He and Joey slid inside, grabbed the cold-water pipes, and felt around to get footholds. They then eased themselves down to the basement.

Moving as quietly as they could, the two boys found a basement exit. Before slipping out, they searched the ground outside the building to find a rock they could use to prop the door open. They then ran across a small yard to the shadow of Waverley Hall, where staff lived on the upper floors. A few trees sheltered them for another hundred feet or so, but they had to make the last half of the journey to the bakery across a wide-open space. They did this unseen, then found the unlocked basement window, squeezed in, dropped down to a metal counter, and then hit the floor.

Inside the bakery, Joey and Freddie quickly found the cakes on large rolling metal racks, which were waiting to be wheeled onto a delivery truck and taken to various buildings. Each of the cakes filled a two-by-three-foot pan. They were so big that the boys could only handle one. Joey slid it out and followed Freddie to the window. He waited as Freddie hopped onto the counter and squeezed out the window. He then got up on the counter himself and handed the cake out.

Back at the dorm, Eric stood at a window watching until he saw the shadowy figures of the cake burglars making their way back across the darkened campus. Their prize was so big they would probably have to come right up the central stairway. If the night attendant was in the office by the front door, they could do this without being detected. But as Eric went out into the hallway to greet them, he heard the door to the office open. The attendant was coming.

Dashing back into the ward, Eric grabbed one of the heavy wooden settees and dragged it out the door. He then picked it up, carried it to the top of the stairs, and with a heave threw it down the stairwell. It hit the landing with a crash that echoed to the roof and back down. The settee then bounced down the stairs, crashing into a wall and then disappearing from Eric's view. The last sound he heard was the attendant running into the office and slamming the door.

Moments later, Joey and Freddie rushed upstairs with the cake. Eric told them what had happened. They knew that the attendant would call for a security officer, who would likely take a long time getting to the dorm. Inside the ward, they put the cake on one of the beds left empty by a Home Boy and dug in with their hands.

Without milk or water, the boys had a little trouble choking down the spicy cake. When they tried to talk, their words were muffled by the cake, and this made them start laughing and spitting spicy clumps out onto the floor. One boy threw his piece of cake at another, but Joey quickly put a stop to the food fight by telling them to save the cake for the others. Freddie kept telling them to hurry up, because he knew that a security guard was on his way.

By the time the security officer got to the BD and helped the attendant drag the settee back up the stairs, the boys had already gone to the annex and washed the stickiness off their hands. The pan had been dropped down the ventilator shaft, and both Joey and Freddie had shed their pilfered clothes and gotten back into nightclothes. When the light was thrown on and the two men demanded to know how the settee had found its way down the stairwell, the boys refused to answer. The men muttered threats and then left.[9]

* * *

The spice-cake theft was a victory over Fernald, but in Freddie's mind, escaping still represented the best way for a State Boy to assert himself. It was a way to declare that he didn't belong in the school. Freddie didn't think about what he might do if he ever actually got out, or whether he might be able to survive on his own. He was just determined to do it.

This lack of serious planning was obvious on the night Freddie made his first run for freedom. It was February 10, 1953. He had recently turned thirteen. That evening, after lights went out, Freddie lay in his bed anticipating what he was going to do.

Since he had once been caught in the ventilator shaft, Freddie chose this time to open a window, reach outside, grab a heavy drainpipe, and hoist himself out. As he did this, his foot searched for a piece of ornamental brick, which gave him a little step. He took one hand off the pipe and quietly closed the window. Then he steadied himself. He was breathing hard, and the cold air froze every exhalation into a puff of white vapor. Though he was perched on the outside of the building, two stories off the ground, Freddie stayed calm. He was strong, and his wiry body was so light that it was easy for him to hang on. After a moment, he began shimmying down. He went quickly, because the frozen pipe was making his fingers numb. In seconds, he had lowered himself enough to drop to the ground.

Once down, Freddie looked up to check for an attendant's face in the window. Seeing no one, he quickly turned toward the east and the lights of Trapelo Road. His footsteps made deep tracks in the heavy snow as he walked a few hundred yards to the edge of the institution's property. Having never been off the grounds alone, Freddie had no idea where he was going. He was unsure about where the road would lead as he turned right and scuffled south to a shopping district called Waverley Square. None of the stores were open, not even the corner gas station. He was freezing cold, and the wind chapped his face raw. He huddled in a doorway and stamped his feet for warmth. Running away was beginning to feel like a bad idea. He was a scrawny kid who still weighed less than ninety pounds and looked like he might be just nine or ten. When two police officers on patrol in their car stopped to talk to him, Freddie didn't move. He was too cold and too scared to run. One of the officers got out, opened the backdoor of the car, and told him to get in. Once he was inside, where the car's heater began to thaw him out, Freddie would only tell the officers that he was cold but otherwise all right.

At their precinct house, the police checked for reports of missing children and, finding none, immediately called the Fernald operator. She told them that no escapes had been discovered, took down a description of the boy they were holding, and then called his likely home, the Boys Dormitory. There, attendants armed with flashlights went through the wards, checking the beds, and failed to discover that Freddie's bed contained a body-double made of laundry. A staff report on the incident explained what happened next:

> The boy was brought back here and turned out to be Freddie Boyce, who had been in his bed at 11 o'clock. Check was made and disclosed his dungarees, shoes and bathrobe under his mattress.

When the police delivered Freddie to the Boys Dormitory, he seemed so shaken and remorseful that he was spared punishment in Ward 22. Attendants confined him to the ward for a few days, but warned him that this was the last time he would be treated so gently. After all, he had risked his life by running away on a frigid winter night, and he had made them look incompetent.[10]

Even as he learned how to elude adults, Freddie was finding the society of boys in the ward more challenging. The epithet he had received from James McGinn—nigger boy—had been adopted by some of the Big Shots. They enjoyed using it to goad Freddie, who wasn't entirely sure what the word meant but knew it was intended to shame him. He could feel the anger rise in his body until it became a hot flush. Then he would attack. Usually, he would wrestle his tormentor to the floor before one or another of the attendants would come and break it up.

As the boys reached ages eleven, twelve, and thirteen, much of what happened in the BD seemed to become more serious. Instead of wrestling, there were now fistfights. And instead of one boy confronting another, now a dozen got into a brawl. Sometimes a boy charged with instigating a fight, or defying an attendant, was referred to a staff psychiatrist for a consultation. Most of these doctors would limit their intervention to a lecture and perhaps the threat of lobotomy or shock treatment at the Bridgewater State Hospital for the Criminally Insane, and boys lived in terror that they would meet such a fate. But one, Dr. Fred Dowling, became known as "the toilet doctor" because he would twist a towel around a boy's head, drag him to the bathroom, and force his face into a toilet bowl.

Attendants contributed to the escalating violence. At one meal, McGinn hit Joey so hard with the ladle he used to find "hollow heads" that the boy's scalp split open and blood began to gush. McGinn took him to Thom Hospital, where he explained that Joey had fallen on the edge of a barrel.

At night, once he ordered the boys to face the fire escape and go to sleep, McGinn demanded complete silence. If he heard talking and no one confessed to it, he would flick on the lights and make a ritual out of calling the boys, one by one, to lean over a table and accept a few whacks with a wooden hanger.

McGinn was not the only one who seemed to use more violence. Several attendants would identify a boy to be the victim and send all the others to jump on him. They called this "pig piling." In other cases, an attendant might use a tough boy to beat up someone he wanted to discipline. The other boys in the ward gathered to egg them on.

Phyllis L'Antiqua, one of the more strict women attendants, used the pig pile on occasion, but she preferred to set up regular boxing matches for two. She often paired Joey, who annoyed her, with one of the Big Shots, anticipating that they would give him a beating. But over time Joey became tougher and more skilled with his fists. When he began turning these supposed mismatches into bloody draws, she stopped picking him for battles.

The State Boys would remember L'Antiqua for one other unusual method she employed to control them. She would quietly approach a boy who was taunting another, or simply talking too loud, and suddenly slap him in the face and box his ears. The shock of surprise was as bad as the sting from her hand, and the boys—who called her "slap happy"—quickly began to hate her for this humiliating tactic. But as much as they despised the organized fights and the slapping, at least the boys of the BD were able to talk about it among themselves. Other incidents, especially the sexual abuse, were so shameful that many of those who were victims chose to keep them secret rather than admit what had been done to them. Later in life, however, a few would tell the truth in order to make the extent of the horror at Fernald clear to others.

Joey Almeida was first sexually assaulted at Fernald when he was eleven years old. He had awakened sometime after 2 A.M. and gotten out of bed to use the toilet. As he turned away from the toilet, the floor's lone attendant loomed in the doorway.

"He asked if I wanted a blow job," Joey recalled decades later. "He said it like he was asking for a match to light a cigarette. Like it was a regular thing."

Joey didn't answer. He had no idea what a "blow job" was. Then the man offered him some candy. He also said he might be able to help Joey get out of Fernald, soon, if only he cooperated.

The attendant knelt before Joey, who closed his eyes tight. He pulled down Joey's underpants, lifted his nightshirt, and began fondling him. Eventually, the man dropped his own trousers and, as he took Joey's penis in his mouth, began masturbating.

When the attendant was finished, he gave Joey a chocolate bar and told him to go back to bed. As Joey walked out of the bathroom, the man called after him, telling him to "just forget about what happened here."

For weeks, Joey had trouble focusing on anything else. He seethed with anger. One day at lunch, when Phyllis L'Antiqua crept up to him and slapped the side of his head for talking in line, he impulsively wheeled and hit her back, right across the face. Joey's response to L'Antiqua's slap was not the most surprising element of this incident. More startling was the way that L'Antiqua's supervisor, Dr. George

Cox, handled the aftermath. He questioned Joey and, when he found out the attendant struck first, let him go without punishment.

Not long afterward, Joey's schoolteacher, Lawrence Gomes, noticed that he seemed especially fidgety. Gomes was a new young teacher who, like Kenneth Bilodeau, approached his students with respect and sympathy. Joey told him what had happened. He referred Joey to psychologist Catherine Chipman, who preserved the assault in her notes.

"He seemed quite restless and admitted that he was nervous to talk about it because it made him feel dirty and the other boys would call him gay boy," she wrote. Chipman told Joey that by telling the story he could help the other boys avoid the same kind of assault. She also promised to protect him from reprisal by the attendant. He then agreed to talk, and revealed what had happened. In her notes, Chipman included details that offer some sense of Joey's calm demeanor.

"During the interview Joseph told the worker that he was going into the bathroom at about 2 A.M. I asked how he was sure of the time? He told worker that he was able to tell time, that he wasn't as retarded as the doctors think, and every ward has a big clock on the wall . . . worker was impressed with Joseph's ability to carry on a conversation. He seemed entirely to know what he was talking about. I feel this boy's story is quite credible and true.

"This is not the first time I've heard of this kind of thing happening. I feel this matter should be taken with the utmost concern as this could have a great effect on any of these boys later in life. For the safety of all our patients there should be an investigation into any alleged sexual assault reports made, in order to assure this kind of thing will no longer occur at this institution." If any action was taken in response to Chipman's report, no note of it appears in Joey's file. Likewise, no record of an inquiry appears in the superintendent's log from this period.[11]

With attendants organizing fights by day and sodomizing them by night, running away became increasingly attractive to the boys despite the risks. Undeterred by his first experience, Freddie tried again. This time, he simply walked out the door of the BD at 3:30 in the afternoon and strode toward the schoolhouse. He went in the front door of the building, but then slipped out a side door and dashed for a wooded area on the west side of the school property. From there he stumbled down a hillside and to a road that led toward the city of Waltham.

It was the last day of February, and darkness fell suddenly. Again lacking a plan, Freddie wandered the streets until he felt cold and frightened. This took less than two hours. He then retraced his escape route, thinking he would slip into the dorm and tell no one where he had been. It almost worked. Freddie got inside the door and just reached the stairway to the second floor when McGinn nabbed him from behind.

When McGinn wrapped his strong hand around Freddie's arm, it came as such a surprise that Freddie was unable to make up a story about where he had been. McGinn seemed almost pleased to hear the boy had tried to run away again. He surely knew that this second offense meant a punishment much worse than a few days confined to the ward. Freddie was going to be locked up.

Twenty-Two occupied the second floor of East Dowling Hall, which was adjacent to the Administration Building. Officially, it was the place where males who might be dangerous to others or themselves could be held securely. In practice, it was Fernald's very own jailhouse, a place for punishment, not protection. All the boys feared going there. Only two boys from the ward had actually seen the inside of the place. The first, Eric Johnson, had downplayed his experience. The second, Jimmy Croteau, had made the place sound scary. Jimmy had been sent to 22 for flirting with girls during a church service at Howe Hall. When he came back, his head was shaved and he told vivid stories about older boys and attendants who had beaten him, stolen his food, and denied him use of the toilet.

Jimmy's stories filled Freddie's mind as he was locked in one of the cells at 22. The room, eight by twelve feet, had a mattress and no other furnishings. Light came from a single bulb that hung overhead, and from a window, guarded with a heavy metal screen, that overlooked the parking lot. He sat down on the mattress and pressed himself into a corner where he could stare at the door and the tiny opening in it, perhaps twelve square inches, that allowed attendants to check on him without coming in.

That evening, when he was allowed into the day room for a meal, Freddie noticed the size and strength of some of the other young men in Ward 22, considered their swaggering attitudes, and felt almost grateful for the protection of his cell. He returned to it, waited as the light was turned off from outside, and then curled up in a ball to sleep. This assignment to 22 was intended as a warning. It would be over in less than twenty-four hours. He would be able to go back to the ward and report to the others that it was a scary place, as they all suspected, but he had not been beaten by the attendants, and the other boys there had left him alone.

Bad as it was, Fred's punishment could have been much worse. Residents of another institution, New York's Letchworth Village, were subjected to electroshock treatment for their "episodes of excitement." Some were given two sets of shocks, every other day, for three weeks. Doctors there reported improved behavior in most, but the results were temporary.[12]

FIVE

A few weeks after Freddie's visit to Ward 22, a young couple brought their seven-year-old son to Fernald for an appointment with Clemens Benda. In his years at Fernald, Dr. Benda had built his reputation to the point where he was the newly elected president of the American Association of Neuropathologists. His name was known by referring physicians around the world, and getting a few minutes with him was impossible for parents whose child had ordinary problems. In this case, though, the doctor was eager to see a new patient. He had been told the boy was a schizophrenic with an IQ of 185, just seven points shy of Einstein's estimated score.

Before he met the family face-to-face, Benda hid behind the door to his exam room and listened to them interact. In the waiting room, the boy pointed out every object that caught his eye and asked about each one in a loud, excited way. It was a stream-of-consciousness process, however, not a conversation. He would ask a question about a lamp or the picture on the cover of a magazine but wouldn't wait for the answer before asking another.

When Benda finally brought the family in for their meeting, the boy continued to talk and to touch everything he saw, ignoring his parents and the doctor. As he watched, Benda began to suspect that he was suffering from profound hyperactivity caused, perhaps, by brain damage resulting from an unrecognized bout of encephalitis. Eventually, he asked the parents to leave and focused on the boy alone.

"When I had him alone and started to test him in a playful manner, he gradually became more responsive and cooperative and revealed an amazing knowledge and ability in giving definitions, reading, and verbal expressions," Benda wrote a few years later.

The boy had brought with him an advanced high school physics textbook. He sat with Benda and discussed the contents of the book, explaining to him in clear terms the meaning of the words and the concepts they spelled out.

Benda determined that the child was not schizophrenic, but he couldn't offer his parents any cure for his hyperactivity. In a way, the visit did more for the doctor than it did for the patient. Benda found the case fascinating because it confirmed his notion that extremes in intelligence—whether very low or very high—might be caused by "uncoordinated development" of the brain. Retarded children suffer a lag in intellectual development, but may acquire good emotional skills. Similarly, some child geniuses demonstrate raw intelligence "that far outdistanced the development of the rest of the mind," he noted.[1]

One year after Benda saw him, the boy with the high IQ was the subject of a pitying article in *Life* magazine titled "A Lonely Little Genius." The piece revealed that eight-year-old Brian Van Dale of Lincoln, Rhode Island, possessed astounding academic abilities. But the main theme of the report was his isolation. Crediting a family physician with the idea, the magazine concluded that the child was a freak, a victim of excess intelligence who was "entirely at ease with his stuffed bear and building blocks because unlike people, they do not interrupt his thinking." Three of the pictures accompanying the text supported the thesis. One showed the boy alone at lunch. Another caught him huddled in the corner of a bus. The third showed him lost in thought during a classroom lecture.

Life's interest in Brian Van Dale reflected both the public's fascination with, and suspicion of, intelligence in general and high IQ in particular. The postwar boom in the development of new technologies had spurred politicians to call for better schools to prepare future scientists and engineers who would make America great. Much media attention was paid to the rigors of education in the Soviet Union, where officials showed visiting journalists high-achieving students whose high school courses were as tough as what Americans encountered at university.

American schools seemed soft in comparison, and alarmists declared a crisis. An antidote was offered in accounts suggesting that intelligence could be increased from the moment of birth if only parents used the right methods. *Science Digest* linked a mother's nutrition to a baby's IQ, and *Coronet* magazine advised parents on "How to Raise Your Child's IQ."[2]

Among professionals, IQ was the subject of a perennial nature-or-nurture debate. In the 1940s and 1950s, the dominant voice in this debate was probably Sir Cyril Burt of London's University College. In a series of papers and books with titles such as *The Backward Child* and *The Subnormal Mind,* Burt reported that identical twins who had been raised apart scored the same on intelligence tests, proving that IQ is determined by heredity. He proposed the use of a national intelligence test for eleven-year-olds. The purpose would be to identify a small group to receive advanced education. A highly persuasive speaker, Burt saw his idea adopted and then put to work across Great Britain.

Though Cyril Burt had many adherents, anyone looking for contrary evidence could find it. In 1949, for example, the *Journal of Genetic Psychology* published a long-term study of one hundred children who had poverty-stricken, low-IQ parents and who had been adopted into stable homes. The paper reported that the children, who were raised apart from their birth parents, averaged twenty-one points higher on IQ tests than their mothers. The authors credited their environment—all the children had been taken in by middle-class families with access to good schools.[3]

Further evidence for the malleability of IQ would have been easy to find in the records of Fernald's State Boys. The trouble was that many, if not all of them, had apparently begun to lose points under the state's care. Freddie Boyce, who scored 79 in 1951, dropped to 77 in 1952 and 74 in 1953. Albert Gagne had a 65 IQ when admitted to the institution in 1948. Five years later, it would be measured at 60. In this same period, Albert's brother Robert would suffer an even bigger drop in IQ, from 75 to 66.

The declining IQ scores received no comment in any of the boys' records. No matter what the popular press was saying, at the Fernald School the prevailing notion, reinforced by principal Mildred Brazier, was that a child's intelligence was permanently fixed, perhaps at birth. Though he was kindly and optimistic, even Kenneth Bilodeau assumed that his students had "only so much gray matter" and were capable only of limited progress. There were moments, like the time his boys made a ten-by-ten-foot relief map of Massachusetts, when Bilodeau saw real learning take place. But he associated this with a kind of training, not with an overall improvement in intelligence.[4]

At Fernald, those teachers who believed that with enough time they could actually raise an individual child's IQ became discouraged by the number of students that were being jammed into their classrooms. Rose Terry, who began work at Fernald's school three years before Bilodeau, was expected to handle physical education for nearly 1,000 children. She spent her first few weeks on the job "in shock, trying not to cry." Though the number of students made her job impossible, Terry was more distressed by what she learned occurred in the dorms. She often saw a student gain self-confidence and coordination in her class, only to lose it to the emotional stress caused by conditions in the wards.

The staffing problems at Fernald and other state schools in Massachusetts had reached a point of crisis, and officials recruited a group of national experts to review the situation. They concluded that the state needed to hire thousands of employees and raise salaries substantially, just to meet minimum standards of care. Though the problems were presented as an emergency, no action was taken in response to this report, and this meant that Fernald's administrators continued to have trouble simply filling existing vacancies.

"They hired a lot of drunks and bums, people who should never work with anything but inanimate objects," recalled Rose Terry at the age of seventy-seven. "Some of the more sadistic ones would get a gang of kids and have them beat a child that was a problem for them. I would see the kid, and the bruises. He might want to squeal, but usually he was too afraid. You couldn't do much then, which was hard to accept. I had to shut my eyes to a lot of things. A lot of us did that— we just shut our eyes—because you couldn't go on if you got upset all the time."

Much of what confronted a young teacher at Fernald required some adjustment. For example, every once in a while Terry and her colleagues were required to attend a lecture at the Southard Laboratory. There, Dr. Benda displayed slides of brain tissue and discussed his theories on human development. Terry found Benda to be brilliant, but she thought the slides were grotesque, and she never understood how these sessions were supposed to help her to be a better teacher. No one ever explained the purpose of the lectures, and she eventually concluded they were simply a way for Dr. Benda to practice professional exhibitionism.

Terry was also troubled by the presence at Fernald of hundreds of normal-seeming boys and girls who regressed emotionally and became more dangerous as they spent more time in the institution. "They were street kids, and we knew they never should have been there. A lot of them were angry because they didn't know how or when, or even *if* they could get out. They would try to escape, but they were always found and brought back."[5]

The tension building in the institution contributed to a rash of strange incidents. On an otherwise ordinary day in the Manual Training Building, Richard Williams found a quiet corner and drank a considerable amount of green paint. He vomited much of it, staining his lips, chin, and shirt dark green. Rushed to the hospital, he had his stomach pumped and did survive. At about the same time, a boy named Charlie Hatch took a sheet into the clothes room, where he shut the door and climbed up the cubbies to tie one end around a water pipe that ran along the ceiling. An attendant discovered Charlie before he could hang himself. He untied the sheet, hit Charlie with a broomstick, and sent him out to the settees.

As an adult, Charles Hatch recalled that the depression that drove him to attempt suicide likely began before his time at Fernald. His mother had died before he was five years old. His father then married a woman who did not want her new husband's three children. She first sought to have Charles's older brother Edward put away as a delinquent, but state officials refused. She then asked a Catholic girls' home to take his sister but was again turned down. Finally, she trained Charles for a performance in a psychiatrist's office. Following her coaching, the seven-year-old boy insisted that flames were coming from a painting on the doctor's wall. Weeks later, he was in Fernald.

As a young child Charlie Hatch had believed his exile to Fernald was punishment for misbehaving. "My stepmother had this cat that I liked to pet. One day it got out and got killed by a car. I think she blamed me," he said years later. "She also had this collection of little china cats, in a locked cabinet. I got in there and took some of them and put them in my room. She hit me with a log from the fireplace for that. At Fernald I just figured I was there because I had done too many bad things."[6]

While self-destruction ran through the minds of many boys, others aimed to destroy Fernald. They stole tools from the shops to cut phone lines and used matches left behind by attendants to set a rash of fires. In early 1954, a small blaze was discovered in a dormitory called the East Building. Weeks later, a brush fire erupted near a building called Tarbell Hall. Two days after Christmas, a patient named Ernest Colley hid behind some curtains in the day room when the boys in Ward 22 were taken to wash prior to dinner. Attendants later discovered Colley breaking windows while a Christmas tree in the ward blazed like a torch. Colley had been upset because he had received no presents or visits and had not been allowed out of the prison ward for the holiday.

No fires were set at the Boys Dormitory, and most of the residents there were too young to stage a successful escape. Only Freddie and a handful of others tried. And even for them, the attempts were more a child's impulse, or adventure, than a carefully planned run for freedom.

Before dawn one summer morning, Freddie and Joey Almeida escaped together, slipping out a first-floor window and racing like greyhounds across the Fernald campus. Once off the grounds, they wandered into the nearby town of Belmont. Joey, who had done his share of shoplifting as a little boy in Cambridge, told Freddie to wait for him outside a small supermarket. He grabbed a grocery cart and walked the aisles. He put some cookies and a cake in the wagon, strolled through the store for a few more minutes, then went past an unattended cash register and out the door.

Apparently, someone in the store had noticed the boy in state clothes and called the police. Though Joey made it outside, two officers rolled up in a squad car just as he greeted Freddie. Seeing the police, the boys ran, leaving their goodies behind.

Powered by fear and adrenaline, Freddie and Joey sprinted into a residential neighborhood behind the store, with the police cruiser in pursuit. Knowing they couldn't outrun the patrol car on the streets, the boys cut between two houses and began running from yard to yard. A little game of catch-us-if-you-can ensued, with the police driving around the neighborhood, changing directions every time they got a glimpse of the boys as they dashed between houses or ran across one of the streets.

Finally, Joey and Freddie crossed in front of the patrol car and ran into an undeveloped tract of woods. The officers parked their car and chased them on foot. By the time they caught up with the boys, they were trapped in a freshwater marsh. Joey stood on dry land. Freddie was ankle-deep in mud, contemplating a plunge through the reeds. He changed his mind when the officers told him he would likely fall into some deep water. Freddie didn't know how to swim.

Though they didn't leave any bruises or cuts, the officers were not exactly gentle as they yanked Freddie and Joey through the woods to their vehicle. Once again, the boys were taken to the police station. After a call to Fernald, the police then took them directly to Ward 22.

This time, Joey and Freddie stayed in their cells for several days of isolation. When they finally returned to the Boys Dorm, they told the other boys of their attempts to steal food and the chase through the neighborhood. With each retelling, Joey and Freddie sounded stronger, braver, and more competent. When they talked to each other about their achievement, they began to believe they could eventually make it on their own, on the outside. It might not happen on their next escape, or even the one after that. But they believed that eventually they would run away and never be caught.[7]

Too old and too big for the BD, in the spring of 1954 Freddie was transferred to the Boys Home. Located on a hilltop that overlooked the entire campus, the BH was one of the oldest buildings at Fernald. It had the same red brick walls and slate roof as the other buildings, but it was more run-down. The floors sagged, many of the faucets and toilets didn't work, and the heat was even less reliable than that at the BD.

Some of Freddie's friends had already gone from the BD to the BH, and his pal Joey Almeida soon followed. What they found there was a dormitory where the residents were as young as fourteen and as old as fifty. The oldest man in the BH was one of the school's few black residents, Charlie Dolphus. A squat and muscular man, Charlie had close-cropped hair that was graying and thinning. Though he was quiet, his presence was a constant reminder to the others that what the attendants said about keeping them locked up for life was possible.

The attendants treated Charlie with some deference. They allowed him to smoke cigars when he was outside and to walk the campus unaccompanied. Charlie was too big to bully. The same was true for Freddie and his peers. Though they still fought with each other, and some continued to be sexually aggressive, at the BH the boys no longer had to worry about being attacked by the staff. They were also permitted to have possessions, including better clothes and books, which they kept in lockers secured by padlocks.

Everything about the BH was more grown-up. A few of the oldest men, who lived in a special ward on the top floor, held jobs off the grounds. Everyone else was assigned work within the institution. Freddie worked for the groundskeeping crew. Joey was temporarily transferred from the kitchen to Dr. Benda's laboratory. He swept floors and washed test tubes, trays, and other items in a basement that seemed like a house of horrors. Along with an autopsy table and rows of stainless-steel instruments, it held an extensive collection of preserved specimens. Huge glass jars filled shelves that lined the walls. Some of these jars contained fetuses or the tiny bodies of deceased infants. Others held spinal cords, brains, or pieces of brains, floating in a preservative. Seeing these specimens reminded Joey that in the BD the attendants used to say, "You better behave or you'll wind up in one of Dr. Benda's pickle jars." He had always wondered what that expression meant.

The adults who worked in the lab seemed comfortable amid the jars, so Joey copied their attitude. He focused his eyes on the floor, desktops, and other people. Gradually, he trained himself to almost forget that he was surrounded by human remains.

Joey's newly developed ability to ignore the grotesque made it possible for him to accept a job promotion. He was enlisted to operate a precision slicing machine, which he used to carve wafer-thin sections from preserved brains. Joey thought the brains were like rubbery cauliflower.

The brains came from the bodies of deceased patients, including severely disabled infants who were being admitted in significant numbers to Thom Hospital. There they had received nursing care until they died. In most cases, parents permitted autopsies, and while Benda was glad to have the material, it created a backlog for the lab staff.

Fernald was not the only institution serving larger numbers of dying patients. With many physicians advising parents to unburden themselves of infants with disabilities, the number of severely handicapped newborns enrolled at state schools across the nation climbed steadily in the 1950s. In many states, new buildings were built to accommodate them, and more highly skilled nurses were hired for their care. But neither nurses nor antibiotics could keep all of these children alive. At Fernald, the presence of these very sick babies pushed the number of resident deaths to 48 in the year 1954. This meant a crush of work in Dr. Benda's laboratory. Joey was relieved when he was eventually transferred back to his old job at the bakery. Albert Gagne replaced Joey at the laboratory and remained there for years, assisting in almost every activity, including autopsies.

While Albert enjoyed the lab, especially the people there, no one was happier in his job than a recent arrival at the BH named Arthur Donovan. One of seven children born to an alcoholic mother, Arthur had been in foster homes and reform school until age fifteen, when he was tested and admitted to Fernald. At the state school, he was assigned to work in the hospital as a nurse's helper. There he was trusted with the most fragile newborns, some with heads swollen to enormous size, causing severe brain damage. Others had deformities such as missing limbs or even microcephaly, a condition in which the skull and brain fail to develop.

Although many people, including parents, recoiled at the sight of these children, Arthur loved them. He eagerly fed, diapered, bathed, and dressed them, but he enjoyed most the time he spent with them in a rocking chair. Warm, soft, and sweet-smelling, these babies felt good in his arms. In the time he spent with them, he forgot that he was confined to Fernald and felt only the pleasure of giving attention and comfort.

Few jobs at Fernald offered the kind of emotional rewards that Arthur Donovan felt at Thom Hospital. But work assignments did make most of the boys at the BH feel a little more grown-up. The extra privileges they received had the same effect. They got later bed times, more television, and every so often the school staff would put on a dance. At first, males and females were segregated. As time went on, and Rose Terry implored Mildred Brazier to loosen the rules, the chaperones became less strict and the sexes began to mingle.[8]

By the time they were in the BH, each of the State Boys had picked out a "girlfriend." Since they had no chance to meet girls and talk to them in private, these selections were made from afar, and almost at random. Freddie chose a blond-haired, blue-eyed girl named Margaret Burney and announced that she was his. Every day, he would stand by a window in the Manual Training Building and watch for her to leave the schoolhouse and walk back to the Girls Home. When he got up the nerve, he wrote the words "I love you, signed Freddie" onto a tiny scrap of paper, rolled it up, and tucked it into the barrel of a ballpoint pen. He then threw the pen out the window. It landed at Margaret's feet, but an attendant scooped it up. Weeks later, Freddie saw a teacher with the pen. "Nice note," he told Freddie.

In their long discussions about girls, the boys of the BH generally agreed that it was better to have a girlfriend from inside Fernald than one from outside. Fernald girls would understand them. Outside girls would not.

On occasional Friday or Saturday nights, some of the older boys sneaked out of the BH—or took a detour as they returned from a movie at Howe Hall—and headed for the open basement windows that let the steamy air out of the girls' shower room. Sometimes, as they crowded around the windows, they managed to get a glimpse of what they hoped to see before a girl screamed. Once, the girls even cooperated, breaking into a giggling, screaming, naked dance before attendants rushed into the shower room and chased away the Peeping Toms.

Though a peek into the shower room was enough to satisfy most of the boys at the BH, a few were determined to get closer to their girls. The boldest sneaked through the utility tunnels that connected the campus buildings to reach the Girls Dorm. They would meet a girl at a designated spot, usually in the basement, exchange a few kisses, attempt to grope her breasts, and then flee.

These kinds of adventures became fond memories for the State Boys, who would recall them long after they became adults. They would speak more reluctantly about sexual incidents involving staff members of the opposite sex and female college students who came to Fernald for internships in training.

Under the Fernald rules, males could not work in the girls' wards, but females could oversee boys. The policy assumed that women were not sexually dangerous. In fact, a few women did use their position to manipulate boys into encounters they long remembered as shameful and humiliating. As an adult, Robert Gagne recalled that a group of women attendants sometimes had one of the more disabled boys take off his clothes, get down on the floor, crawl around, and bark like a dog. "He was very mature physically," said Robert, "and I think they liked looking at him."

On other occasions, women attendants stripped to their underwear, got in the showers with boys, and performed sexual acts on them. One summer a group of female college students, who had gotten internship assignments as attendants in the Boys Home, brought groups of boys into the annex, where they coaxed them into masturbating. They laughed and applauded and gave candy and cigarettes to the boys who reached orgasm first.

Though not widely recognized outside the institution, the problem of female attendants abusing male residents was long-running at Fernald and likely existed at other such schools. As early as 1922, a notice to women employees at Fernald reminded them "not to make love to the boys in their charge." Each attendant was required to sign a copy of this notice, which also observed that such practices "always create harm."[9]

The harm, for the boys of Robert Gagne's era, came from the confusion and embarrassment they felt about their sexual encounters with women staff. They liked some of the attention they received, enjoyed the sexual release, and became infatuated with the women. But they also understood that what they were doing was wrong, that they would be punished if caught, and that they were being manipulated by attendants who held power over their lives.

Equally troubling were the relationships that developed among the boys themselves. Several older boys in the BH were what they all came to term "night crawlers." After the lights were turned off, they would get into a younger or smaller boy's bed, threaten him into silence, and then teach him to engage in mutual masturbation and oral sex. Often a relationship would evolve, as an older boy claimed the younger one as a sort of pet. During the day, the younger boy would receive favors and protection. At night, he would perform sexually.

These pairings were recognized by the boys in the wards, and since they always involved a Big Shot who could be menacing, they were accepted without much comment. Attendants did nothing to interfere, believing that the behavior was none of their business. Like prison guards who tolerate sex among inmates, some of the adults at Fernald thought that the affection and protection that were part of these relationships might be beneficial for those involved and help calm the entire ward.

When he first came to the BH, Joey Almeida had belonged, in this way, to a night crawler named Willie Adams.* He was three years older than Joey, and a dominant figure in the dorm. Joey didn't feel ashamed of the relationship. As far as he could see, it was common practice in the dorm, and Willie spared him from having to deal with multiple abusers. It went on for a few months, and then Willie let Joey out of the arrangement. By this time, Joey had adjusted to the BH and didn't need protection.[10]

Tough boys like Willie Adams ruled the Boys Home more with intimidation than with violence. At the BH, the age and size difference between the Big Shots and the others was far more pronounced than it had been in the Boys Dormitory. These young men had the strength to kill and sometimes acted as if they were willing to do it. After one escape, a Big Shot named Earl Badgett even brought a handgun back to the BH, which an attendant discovered among his clothes.[11]

Often the dominant boys were those whose families remained in contact, sending letters and visiting on most Company Sundays. Bad as their circumstances were, these boys drew strength from the attention. It gave them a sense of being superior to the others and fed their confidence. They rarely used physical force to get what they wanted. Threats were enough to get other boys to give up food and clothes, and to provide sexual services.

Albert Gagne, who was still the smallest boy in his group, suffered at the hands of these boys and eventually developed panic attacks. These often occurred at night, when he was in bed, and the events of the day swirled in his mind. There was much to fear at Fernald besides the Big Shots, including attendants who were both violent and sexually predatory. One summer night, the worry and panic built in Albert until he began to feel like he couldn't breathe. With no place to go, Albert didn't intend to run away. But he was desperate for fresh air. He lay in his bed, waiting for the attendant on duty to make her ten-thirty bed check. She came into the ward, walked up and down the rows of beds, and then departed for the upper floors.

As she left, Albert got out of his bed. The others, exhausted from a hot day's work in the huge gardens, slept soundly. Albert moved quietly to the door, looked and saw no one at the desk in the hallway, and slipped out.

A full moon shone on the barefoot boy whose nightshirt clung to his legs as he ran down the hill toward the baseball diamond and the fields beyond. He crossed the road that brought cars from Trapelo Road onto the Fernald grounds and then dashed behind the backstop. There, the slope of the hill that led to Dr. Farrell's house and a small grove of trees offered him a good hiding place, and he paused for a moment to catch his breath. He then walked down the third base line, into left field, and disappeared into rows of corn high enough to hide him.

Once he was deep inside the cornfield, Albert sat down and listened to his heart beating in his eardrums. He stared up at the full moon and began to relax. As the pounding in his chest subsided, he lay back with his hands behind his head and stared straight up at the sky. He heard crickets, and a breeze blowing through the stalks of corn. In a few minutes, he would get up, leave the cornfield, and sneak back into the BH. But until then, he would feel calm, and safe.[12]

* * *

With no visitors, and not even the image of a family in his mind, Freddie Boyce had little reason to hope for rescue. Nevertheless, he often daydreamed about being adopted. He had never heard of such a thing happening to a Fernald boy or girl. Others laughed when he mentioned the idea, but it was the only event that he could imagine might save him.

Though Freddie was completely unaware of it, a Fernald psychologist had reached the same conclusion. In the fall of 1954, Freddie had been given a new series of intelligence tests. The result had been startling. Instead of continuing the decline recorded in previous exams, his score actually increased a full twelve points, to 88. This number was so high that the tester wrote that "a foster-home placement and placement in a public school system would be of great benefit" to Freddie. More to the point, the psychologist argued that "a prolonged stay here would be detrimental to him."[13]

These notes were written in November 1954, as the state was creating a Division of Special Education to develop community school programs for children with learning disabilities. In a few years, more than a thousand children were enrolled in such "special-ed" programs. This phenomenon was a response to growing demand from parents and to the overcrowding at Fernald and the commonwealth's three other state schools. But it was too late for Freddie. He was already a State Boy. No further mention of moving him out would appear in his records, and he was not told about what the psychologist had recommended.

However, the sensitivities revealed in this one psychological evaluation of Freddie Boyce were suggestive of changing attitudes about institutions. In 1950, a few dozen parents had formed the National Association for Retarded Citizens with the intent to press for improvements in institutions, more community-based programs, and a new, positive public image for their sons and daughters. The organization was especially strong in New York, where Jewish parents who were accustomed to social activism played a leading role. Years later, these founders noted that the assertiveness of returning World War II veterans was also important to the cause. Men who had fought for their country were not going to accept second-class status for their children.

By 1954, the NARC had nearly 30,000 members in more than 400 local affiliates. These parents were so effective at prodding government officials that many states formed commissions to review care of the retarded. Federal spending on these children went up fivefold. President Eisenhower even declared a National Retarded Children's Week in order to fight the stigma associated with retardation.[14]

In Massachusetts, Governor Paul A. Dever seized the political initiative, joining Malcolm Farrell on a TV program titled "New Hope for the Mentally Retarded" and appointing a commission to investigate the state schools. This commission documented conditions of filth, overcrowding, and neglect. It recommended sweeping changes, including construction of new facilities and improvements in staffing.[15]

While the state embarked on the long, slow process of making policy and approving construction, the overcrowding and understaffing at Fernald caused more problems. Malcolm Farrell found himself dealing with a steady number of reports of staff abusing patients. In November 1955, he fired one employee for hitting a resident at Fernald and had to go to court to testify against another who was charged by prosecutors with "unnatural acts" involving a young man at the Templeton Colony.[16]

Though the presence of violent attendants posed real peril, sometimes the mere absence of a worker had more tragic consequences. Some of the State Boys forever remembered an unsupervised walk they took near the school's boundary with a local Girl Scout camp. "This one kid had a toy gun," explained Larry Nutt. "Somebody grabbed it and threw it over the fence. It landed in this pool that they had over there. Well, the kid loved this gun, so he went over the fence to get it. He went right in the water. But he didn't know how to swim. The kid drowned, right there in front of us, and we just stood there not knowing what to do."

Bad as Fernald was, it would have been relatively easy to find similar neglect and abuse in most state schools. At this moment in American history, thousands of retarded children lived in barren rooms, where the stench of urine and feces was so powerful it made visitors vomit. Gunnar Dybwad, onetime director of the National Association for Retarded Citizens, would recall seeing one incontinent patient at a New York school for the retarded lying naked in a box of sawdust. In

Tennessee, a proud superintendent showed him how young children were rolled on rubber mats down a chute to be bathed like hardtops in a car wash. State legislators in Alabama literally cried after touring a decrepit state school there.[17]

As the public received more information about institutions, reform slowly became a popular cause. The innocence of the retarded child made it easy for politicians, parents, and professionals to become her champion. The Kennedy family, through its foundation, became deeply involved in research on ways to improve care. By the 1960s, President Kennedy would begin large federal programs to make their lives better.

Unfortunately, in all the efforts to expose, understand, and then deal with the troubles of state institutions, hardly anything was ever said about those, like the State Boys, who didn't belong in them in the first place. Insiders were well aware of the problem. They even warned of the danger posed by so many angry teens. But with no parents to advocate for them, the plight of these residents was generally overlooked. They were not innocent-looking, harmless-acting children who evoked public pity.

Oblivious to the professional argument that had been made for his release, Freddie focused on those Company Sundays when he spotted Bea Katz and Louis Frankowski out for their walk. He ran to meet them and stayed at Mrs. Katz's side for the length of her visit. She liked talking with Freddie and encouraged him to believe he didn't belong at the institution. But Mrs. Katz didn't have the time, or energy, to help him in the ways that she was helping Louis. With her regular attention, and the Fernald staff finally making an effort to train him, Louis had made significant progress. He had begun to talk, had developed good control of his bodily functions, and was learning to perform some small tasks with his hands. Mrs. Katz's husband, George, joined her in many visits, and he also became devoted to Louis. They had started referring to him as "our moral child."

Mrs. Katz often appeared at Fernald on days other than Company Sundays. A few attendants liked her and appreciated what she was doing. They would open the ward when she came. On some of these visits, she would discover that Louis had been lying for hours abandoned in a bed, or that other residents were harassing him while an attendant read the paper or had walked away for a coffee break.

With the fierceness of a mother, Bea Katz complained about Louis's care to attendants, their supervisors, and administrators. She succeeded in getting him moved out of the depressing, filthy wards where the most disabled patients were warehoused and into one where he would receive more education and training. As months and then years passed, she saw improvements in Louis's speech, his appearance, and his mood.

If Freddie was jealous of the attention that Bea Katz gave to Louis, he didn't show it. But he was, of course, starved for affection. Chatty to the point of being annoying, he would talk to every new person he met, hoping to make a connection with someone who might try to get him out of Fernald. It finally happened in the spring of 1955, with the help of a skunk.

The skunk had been hiding in a drainage trench near Dr. Benda's lab building. Freddie, working with the groundskeepers, was handed a rake and told to clear the trench of old sticks and leaves. He jabbed his rake into the trench and the skunk responded, spraying his leg. As the animal scuttled across the grass, Freddie ran to the Boys Home to wash. But no amount of soap and water could make him clean. The attendants pushed him outside and pointed him to the resource of last resort—Thom Hospital.

The nurse on duty in the emergency department on that day was thirty-six-year-old Mary Mone. She had come to Fernald the year before after working in several hospitals and then trying with her husband to make a go of it with a little grocery store. When the store was defeated by the rise of the new supermarket chains, she returned to nursing. One of her first assignments at Fernald had involved administering hundreds of gamma globulin shots to stave off a hepatitis epidemic. (The outbreak had been started by a young female attendant who had contracted the illness from a veteran receiving treatment for the disease at a nearby Army hospital.)

When Mone had gone into Fernald's dorms to give the gamma globulin, she had been upset by the crowding—the school census was up to 2,434—and by the stench and squalor of the back wards. Knowing how they lived in the dormitories, Nurse Mone sometimes cried for the children she treated for scrapes, bumps, sore throats, and coughs. A twelve-year-old girl in a straightjacket, which Mone demanded be removed before she administered a tranquilizer, would remain fresh in her memory for fifty years. But few children affected her as much as the boy who came reeking of skunk and desperate to be rid of the stink.

Baths of tomato juice—a well-known country remedy—did the job. When the odor was gone, Nurse Mone rinsed Freddie and then let him soak a bit in the tub. It was then that he began to talk, and talk. Nurse Mone, who struggled to keep from laughing at his commentary about the skunk, could see that he was lonely. She asked a few questions and learned that Freddie had no family, received no visitors, and had not spent a single night away from Fernald since his admission. She decided to change that.

In the summer of 1955, Nurse Mone wrote a letter to Malcolm Farrell requesting she be allowed to take Freddie, and Arthur Donovan, whom she had met at Thom Hospital, to her home for a week. Childless at the time, the Mones had the means and the desire to offer the boys a break from institutional life.

To Mary Mone's surprise, Farrell granted her request, with a single stipulation—that the boys not leave the commonwealth. On a Friday afternoon, she took them to her house near the little city of Billerica. That evening, she and her husband, Raymond, took them to their favorite restaurant, a roadhouse with a cluster of overnight cabins called Duca's. Inside, the boys and the Mones sat at one of a dozen-or-so booths and listened to the jukebox. After eating, as Mary and Ray Mone chatted with friends in the restaurant, Arthur and Freddie found the busboy, the owner's son Billy, who was about their age.

Almost fifty years later, William Duca could remember that conversation. "Those kids seemed just like me," he said. "But when they told me where they were from, I almost couldn't believe it." As his new friends left, Billy Duca reflexively said, "If you ever get out, come see me. I'll help you out."

That one visit to the restaurant satisfied the two boys from Fernald, who had been hoping only for a glimpse of the outside world. But Ray and Mary Mone had much more planned. The next day, they took the boys on a car trip to Lake Placid, New York—definitely out of the commonwealth—where they stayed at a little resort, swam in the cold lake, and forgot they were State Boys.

Arthur and Freddie returned to the BH with a stack of tourism flyers from Lake Placid and showed them to all the other boys. Superintendent Farrell either never learned that the boys had been taken out of state, or chose to ignore it. Mary Mone would be allowed to take the boys to her home for several more visits. Freddie became especially attached to her husband, Ray. He let the boy help him build a recreation room, complete with a bar, in the basement of his house. Freddie also helped mow the lawn and did other chores that made him feel like a regular kid.

Ray Mone became the only father figure Freddie would ever know. He loved tagging along as Ray did his woodworking or some yard work. Though they didn't talk about serious things, Ray's presence was deeply comforting to the boy. He grew in Freddie's eyes until he began to think of Ray as an equal to Ted Williams, the ultimate hero in the eyes of every boy at the BH.

At Fernald, Freddie began to talk about the Mones as if they were his family. In Mr. Bilodeau's class, he spoke excitedly about the couple who had allowed him into their home. Then Bilodeau overheard Freddie say that he was certain that the Mones were going to adopt him. Freddie even claimed that Ray Mone had asked him if it would be a good idea if he became Freddie Mone.

This was why the school discouraged staff members from taking children off the grounds. Without any encouragement, the child always developed a kind of crush and became certain he would be rescued and swept into a loving home. Bilodeau was sure that Ray Mone hadn't said what Freddie reported. The teacher was also sure that if he held on to the dream of adoption, Freddie would only be disappointed. The solution, he decided, would be to convince Freddie to reject the Mones before they rejected him.

Bilodeau and Freddie discussed the adoption issue many times. They talked about how Freddie, now a teenager, was practically grown up. They considered the difficulty he might have adjusting to life outside Fernald. Would he know how to live with a family? Could he handle the work in a regular school? What about the kids he would meet on the outside? Gradually, Freddie began to agree that adoption was a bad idea. He agreed that he was too big, too close to being a man, to become someone's little boy.

As he came to accept that he was destined to be an orphan forever, Freddie became more difficult in the dorm. His record from this time shows that he had started to swear and to talk back to attendants. "He resents direction, unless it happens to coincide with his own wishes," wrote school principal Mildred Brazier, who later in her report noted that "Fred," which he was now called, "keenly feels the lack of family ties."

The death of his dream of adoption was one factor in Fred's deteriorating attitude and mood. But there were other things contributing to his anger. Old enough to attend a regular high school, Fred was still stuck with schoolbooks and lessons from the third and fourth grades. Fernald didn't offer anything more advanced, and this frustrated him. Then there was the strange encounter with a long-lost brother.

As far as Fred knew, he was the only child of parents who were long dead. In fact, school officials knew he was one of at least six children. One, a brother George, was also a ward of the state. Never diagnosed as feebleminded, George had been taken in by a private children's aid organization called the Home for Little Wanderers. At about the time when Bilodeau was persuading Fred to drop the adoption idea, George's guardian at the Little Wanderers orphanage decided to bring him to Fernald to see his brother.

Fred was in the gym bouncing a dusty basketball when a teacher named Joe Barnes appeared with another man and a pale, slim boy who looked younger than Fred. The boy wore street shoes, a white shirt, and a tie. Barnes called to Fred. "Get over here, there's someone I want you to meet."

After Fred trotted over, Joe Barnes told him to take a moment to catch his breath and get ready for a surprise. Then he said, "Fred, this is your little brother, George."

Stunned into silence, Fred waited for George to stick out his hand

and say, "Hi." He grabbed the hand, shook it hard, and said, "My brother? For real?" Barnes assured Fred that George was his brother and that he was one year younger than him. He pointed to a little grandstand and suggested they go sit and get acquainted.

In the next half hour, Fred heard how George lived in a cottage with a small group of boys and attended regular schools. Fred tried to explain his own life at Fernald but felt embarrassed by the institution. He was also self-conscious about the way he talked. George sounded much more intelligent.

When it was time for George to go, Barnes allowed Fred to walk him to the car. He stood and waved as the other man drove away with George. Fred returned to the gym and joined a pickup basketball game. Later, when he was in bed, he thought about what the visit meant to him.

From the moment that Fred was first able to wonder about such things, he had believed he was a sort of alien, and the phrase "from another planet" echoed in his mind. The sudden appearance of a brother who looked and sounded a little him made him feel that somewhere there were people who might claim him.

On the ward in the BH, Fred began talking more and more about feeling imprisoned in a place where he had no chance at an education and no discernible future. He would tell Fernald staff that he wanted to be "a scientist, or an astronomer." They would regard these dreams as the "unrealistic compensation" of a child who had yet to accept his station in life.[18]

Believing that he belonged there, and not in Fernald, Fred studied the outside world. His best tool for this work was a crystal radio set, which he built after seeing a plan in a book in Mr. Bilodeau's classroom. He got some copper wire from one of the shops in the Manual Training Building and wrapped it around a discarded toilet tissue roll. He used a bit of quartz he found on the grounds for a crystal, and a bent safety pin became part of his tuner. Pat O'Callaghan, supervisor of the Boys Home, brought him a headphone set to complete the kit.

To make the whole thing work, Fred crawled under his bed—one of the few places where he could find privacy—and bent the antenna wire around a steam pipe. The ground wire he attached to the frame of the bed. After putting on the black headphones, he deftly grazed the set's copper coil with the point of the safety pin to isolate signals from local AM stations. He heard music, news, ball games, in short, the sound of the outside.

The radio was exciting, but the more Fred understood about what he was missing in the world beyond Fernald, the angrier he became. He got into more fights with other boys. He stole clothes, candy, and cigarettes. Though he often protected the more retarded Home Boys from the taunts and threats of others, Fred was not above taking advantage of them in certain ways. In the spring of 1957, now sixteen years old, he tricked a few of them into trading quarters and dollar bills for piles of pennies. There was a purpose in all this. He was planning a reunion with his friend on the outside, Billy Duca.

Before dawn, Fred made a pile of clothes on his bed and threw a sheet over it. He opened a window and swung himself out. His ward was on the first floor, so the drop to the ground was just a few feet. He ran down a hill, toward the town of Waltham, and then made his way along back streets to Waverley Square. At sunrise, he began walking east, holding his thumb out to beg for a ride.

In the 1950s, hitchhiking was common, and it was not unusual for drivers to see lots of teenage boys out early on a summer morning trying to catch a free ride to work. Unlike other times, when he escaped with no destination, Fred was able to tell the drivers who stopped to offer rides that he was headed for Billerica and the highway café. Three rides got him where he wanted to go.

If Billy Duca was surprised to see Fred Boyce come through the door of his parents' restaurant, he didn't show it. Worried that his father would spot Fred and send him back to Fernald, Duca calmly hustled him outside. It was the height of the travel season, and the cabins in back of the restaurant, where Billy might have hidden Fred, were all occupied. Billy showed Fred a secluded place in some nearby woods and then brought him some food and a blanket.

Fred spent three days and nights in the woods. On the third morning, he awoke feeling more thirsty than he had ever felt in his life. (Billy had not brought him enough to drink.) He pulled himself up and walked back to the cabins, taking care to avoid being seen. After listening at walls and peeking into windows, he found that one of the little buildings was empty. He tried the door, found it open, and went inside to find a drink.

Just minutes after Fred got inside, a maid came into the cabin and asked if he was a guest. Unable to fashion a quick lie, he admitted that he wasn't, and she immediately brought him to the restaurant. Billy Duca's father recognized him and, after questioning his son for a moment, telephoned Mary Mone. She came quickly to take Fred to her home, assuring the Ducas that they needn't notify the state school or call the police.

The sight of Mary Mone, and later the warmth of her home, made Fred beg to stay. When her husband, Ray, came home from work, Fred implored him not to call the authorities. But the Mones understood that something more than Mary's job was at stake. By keeping Fred, they would be breaking the law. They couldn't afford the risk. Fred could stay the night, but in the morning he would be going back to Fernald. They promised him, however, that he would not be punished.

The Mones did not have the power to keep their promise. As soon as Fred arrived at the BH, he was transferred to Ward 22. He was taken through the heavy front door, up the stairs, past the day room, and into a cell. The door closed and the lock was turned. This time, Fred spent three days in solitary.[19]

SIX

Ward 22 would never be a permanent solution for the turmoil at the Fernald State School. It just wasn't big enough to hold all the angry boys who, with enrollment climbing to 2,600, were being jammed into already overcrowded dormitories. Fires, false alarms, and break-ins occurred at the storeroom, the bakery, the lab, and the schoolhouse. Most of these incidents were the product of boredom and resentment. Joey Almeida, for one, got a sense of accomplishment and a taste of revenge every time he broke the locks on the vending machines in staff residence halls and stole the coins and the merchandise.

The warehouse, where food, clothing, and other supplies were received and stored, was a favorite target for the angry young men of Fernald. On an almost weekly basis, boys from the BH slipped out of the dorm and crossed the grounds in the darkness. They entered the building through a door or window that one of them had left unlocked after his shift as a warehouse worker. Once inside, they opened cases of food and consumed huge snacks, or stole items from the stacks of used clothes, which had been donated by church groups and others on the outside. None of them felt guilty about what they were doing. The clothes were going to be distributed to them eventually. They were just making sure the attendants didn't take the good stuff first. And like all teenage boys, they were ravenously hungry most of the time. The way they looked at it, if the state didn't feed them adequately, they had a right to steal what they needed.

Fernald staff knew which boys were committing these crimes. But their options for dealing with them were limited. In the past, repeat troublemakers might have been transferred to other institutions, including the Belchertown State School in western Massachusetts and Wrentham to the south. But these facilities were filled to overflowing, too. The place of last resort, the prison called Bridgewater State Hospital for the Criminally Insane, was no longer an easy option. The Massachusetts courts had recently granted new rights to juveniles in state custody, requiring hearings, mandating they get legal counsel, and limiting commitments for evaluation to just thirty-five days.

Other factors made it difficult for Malcolm Farrell to discipline the State Boys. Politicians, including the powerful Kennedy family, were beginning to make the retarded a special cause. And the middle-class parents of the Fernald League were monitoring his every move and would complain to state lawmakers if they didn't like something he did. The league was part of a national phenomenon. All over the country, parents groups were putting superintendents like Farrell in a difficult position. On the one hand, they believed Pearl Buck's message about the value of institutions and wanted an end to waiting lists so their children could be admitted. On the other, they demanded that schools that were already overburdened and understaffed offer much better care and education.[1]

Farrell tried to show the outsiders that he was open-minded and modern. He permitted more volunteers to visit and met frequently with league members. He also began urging an end to the use of the term "feebleminded," suggesting instead "mentally subnormal," which he considered less stigmatizing. Although these small changes may have pleased sensitive parents, they did nothing to stem the discipline problems within the school. Farrell had to know that the defiance and resistance were likely to continue because, as Clemens Benda observed in a memo, the institution itself was the problem:

> Patients are told they are here for good. They lose all hope of ever getting out, being discouraged by other patients. They also rarely see any advantage in showing good behavior. While bad behavior is usually punished, good behavior is always taken for granted and does not bring any merits.[2]

The problems mentioned by Benda would become the main responsibility of psychologists and social workers who were hired as Fernald began to receive new federal funds allocated by the U.S. Congress, which was responding to public concern about mental health. Between 1956 and 1961, money for such programs increased fivefold. The money drew a new generation of better-trained, highly motivated professionals to institutions. The younger staff members increased the use of psychotherapy with residents and would eventually create formal programs for training them to live on the outside.

The training projects would take years to develop, recalled Raymond Pichey, who was a Fernald social worker at that time. Pichey—pronounced Pee-*shay*—transferred to Fernald after a stint at a state psychiatric hospital, where his patients had included the Red Sox ballplayer Jimmy Piersall, made famous by his nervous breakdown. The insular environment at Fernald had dismayed Pichey. With visiting restricted to four hours per month, and much of the staff living on the grounds, the place "was totally detached from normal society." Fernald was so isolated it had developed a genuine subculture, with rules and norms that allowed for the abuse of patients and cover-ups, said Pichey. (According to the Fernald way of thinking, outsiders couldn't understand the challenge to maintaining order in the institution, or that occasional abuses were to be expected, considering the situation.) Pichey was also discouraged by the resistance he encountered as he sought to parole State Boys to live on the outside.

"They were afraid of what they might do in the community, and worried about the institution losing the labor the kids did," said Pichey. "At meetings Mildred Brazier would say something like, 'I saw him staring at a book of matches,' and that would be it for that kid. Sometimes Dr. Farrell would tell me that he wanted to go ahead and parole someone, but he couldn't go against the committee all the time. So we had a lot of kids who should have been out, who just waited, and waited." In the meantime, Fernald's staff needed help to quell the rebellion. It arrived in the form of a pill.[3]

Thorazine, the first of the modern class of tranquilizers called neuroleptics, was used to produce what some researchers called its "lobotomylike" effect. Patients on the drug became compliant and exhibited a silent detachment. At Fernald, some State Boys participated in trials of the drug. They were called into the day room, handed pills, and instructed to swallow them. Some held the tablets in their mouths and later spit them out. Others received placebos that had no effect. But those who swallowed Thorazine during the daily medication calls were suddenly much more docile.

The Thorazine solution was not confined to Fernald. As part of an experiment, doctors at an institution in Washington State put more than forty inmates on the drug. The result was "their halls, formerly often in turmoil, became more peaceful and quiet. Destruction of property and clothing diminished, compensating for the cost of drugs. The attitudes of ward personnel improved, and physical restraint of patients could be eliminated." Similar results were achieved at Vineland in New Jersey, where, it was noted, "all patients now enjoy more pleasant surroundings."[4]

In its time, Thorazine became the dominant drug of its class. But in 1957, scientists were racing to develop other compounds to alter the mind. Boston was a center for this research, and Clemens Benda followed developments with intense interest. He even teamed with Dr. Max Rinkel of the Boston Psychopathic Hospital to conduct an experiment that involved giving a patient lysergic acid diethylamide (LSD) and then monitoring the "artificial psychoses" that developed. Though Benda's involvement was fleeting, Rinkel participated in many LSD trials, including one that appeared to cause the suicide of a patient.[5]

No pill could bring real peace to Fernald, especially when it came to the young men of the BH. Although they had good reason to react to their imprisonment at Fernald, they also found inspiration for rebellion in the style and the attitudes of Marlon Brando, James Dean, and Elvis Presley, whom they saw on television. Newly aware of fashion, the boys often fought over the shirts and pants that were available in dorms. They wanted clothes and hairstyles that made them look like their heroes, and their peers on the outside.

In moments alone, the boys frequently complained to each other about the attendants and recalled incidents of past violence: McGinn's bed-iron torture, L'Antiqua's slaps. They also tried to detect patterns,

clues to basic human nature, in the behavior of the staff. "It was obvious that people changed when they started working at Fernald,"
recalled Fred Boyce. "After a while, the pressure of the place or maybe
the attitudes of the other attendants got to them. They would do
things they never would have done in the beginning. I think it taught
me that anybody can turn bad."

Fred Boyce also saw changes in the behavior of the others in his ward.
Charles Hatch, for example, had changed from a suicidal boy to an
assertive young man. Powerfully built, with thick black hair that he
groomed into a style like Elvis Presley's, he was respected even by the
toughest Big Shots. Hatch's new status as a teenager was a marvel to Fred.

The transformation had started when Hatch was transferred to the
BH from the BD. Soon after his arrival, he was surprised to learn from
Charlie Dolphus, the old man of the ward, that his own father, Edward
Hatch, Sr., had been a State Boy at Fernald from 1913 to 1921.
Dolphus had met the elder Hatch, and told stories about their time
together in the institution. Knowing that his father had firsthand expe-
rience with the institution made Charles Hatch angrier about his fate.
How could his own father have abandoned him to this place, he won-
dered, if he knew what it was like?

Enraged by this, Hatch began to run away and to get into fights.
Banishment to Ward 22 did little to change his attitude. Instead, he
returned to the BH feeling angrier and more rebellious. Finally, instead
of isolating him in 22, the Fernald staff decided to put him in the
North Building, home to more than a hundred severely disabled older
men. They stripped him of his clothes and then dressed him in a blue
jumpsuit that buttoned in the back. Then they walked him down a
hallway toward a large day room. When they unlocked and opened the
door, the boy was hit by the stench of urine and feces. He was pushed
inside, and the door was slammed shut.

In the North Building day room, puddles and piles of human waste
littered the floor, and the room was filled with rhythmic moans, chirps,
and shouts. A crowd of stooped and drugged men, many of whom were
either half-dressed or completely naked, quickly surrounded Charlie
Hatch. Terrified by the men who touched his face, and pawed at his
body, Charlie tried to find a safe corner to sit in. However, the room was

barely furnished, and the few benches that were available were occupied. So he moved slowly around the ward, fending off the patients until they got used to his presence and left him more or less alone.

Exposure to the North Building was punishment enough for any boy from the BH, but the attendants made sure to add some public humiliation to Charlie's experience. They decided to organize the men of the ward, at least those who could follow instructions, for one of their walks outside. Charlie was ordered to be part of the group, so out to the yard he went, in his blue suit, to walk in a big circle with his hands on the shoulders of the man in front of him. When he went back inside, Charlie was brought to a tiled shower area called the Blue Room. He was handed a hose and directed to wash excrement off the bodies of men who were brought in, one by one. Two large, menacing attendants stood watch, to make sure he was thorough. When this chore was finished, he was allowed to get back into his regular clothes and return to the BH.[6]

The ward that Hatch returned to at the BH was tense and unsettled. Attendants were actually afraid of some of the Big Shots, and they constantly negotiated for peace, trading extra food, cigarettes, even a measure of authority, for cooperation. The bargains bought short-term calm but also made the Big Shots bolder.

The situation over at Ward 22, where the tougher young men were collected together, was even more volatile. On the last day of February 1957, a State Boy named Clifford McIntyre, a dominant resident of 22, lost control. An argument with an attendant who had uttered a racial slur—McIntyre was black—escalated quickly. McIntyre began throwing chairs and busting windows in the ward. Though Clifford wasn't big, he was strong, and he was powered by so much adrenaline that five employees were injured as they tried to control him. Eventually, they managed to wrestle him into a seclusion cell, but three days later he escaped. After two weeks on the run, he was picked up by the Boston police, charged with assault for his fight with the attendants, and sent to the state prison in the town of Bridgewater.

These events were recorded by Superintendent Farrell in one of his reports to the Fernald trustees, along with the observation that McIntyre's IQ may have been as high as 84, which made him a "borderline case" for enrollment in the institution. What Farrell didn't note, and may not have known, was that McIntyre had ample reason to exhibit a bit of rage.

A few weeks before McIntyre's outburst, he had had a series of minor disputes with one of the more belligerent attendants at Ward 22. Finally, on an otherwise quiet afternoon, the attendant invited McIntyre for a walk outside. He said he wanted to talk things over. He led McIntyre down a path in the woods to a spot where a group of men, outsiders, jumped from a hiding place and beat him severely. Word of the attack spread quickly among the male inmates. McIntyre recovered quickly but then exploded in the incident that landed him in state prison. After he was gone, the boys left behind wondered if McIntyre might actually have an easier time of it in a correctional facility.[7]

It may have been logical for Malcolm Farrell and his staff to conclude that other aggressive boys would be chastened by the swift and severe action taken against McIntyre. They weren't. Ten days after McIntyre's escape, six young men pushed past two guards to break out of Ward 22. Once they were off the grounds, they met up with five young women who had simultaneously escaped from their dormitory, the Northwest Building. Local police found the group off the grounds, but while the females were quickly caught, the males ran and were gone for days before police in another town nabbed them.

As the eleven inmates were being pursued by the police, Joey Almeida and another boy sneaked out of the Boys Home and broke into the Manual Training Building. They managed to stay there, undetected, for days. One night, they left their hiding place, stole a cigarette machine from a staff dormitory, and brought it back to the Manual Training Building. After breaking it open, they pocketed the contents of the coin box, took dozens of packs of smokes, and then sold the rest to other boys. When Joey and his accomplice were finally discovered, they were armed with spears they had made from knives and broomsticks. They ran, got off the grounds, and had to be tracked down by Boston police. Among the items discovered in their possession when they returned to the BH was a stolen set of passkeys that had allowed them free range of the institution.

Clifford McIntyre, the gang of boys and girls who escaped, and the two who moved into the Manual Training Building would have been enough to strain the resources of Fernald's staff. But in this same time period, three other boys, including Charlie Hatch, escaped for periods of a day or two. As he reflected on fifteen days of mayhem, Malcolm Farrell struggled for a way to explain it to the trustees.

"These episodes are, it is believed, a manifestation of general unrest not only in institutions but in the community," he wrote in his report. In staff meetings "the question was raised as to whether we were short of attendants," he added. "The superintendent agreed that we were, but it was largely due to quotas being set too low."[8]

Almost all American institutions were understaffed, and those workers who were on the rolls were underqualified. This problem had been obvious to the superintendent of a British institution who had made a tour of the country's state schools in early 1957. In a report that was polite but highly critical, D. H. Thomas noted that most American institutions were far too large, too isolated from the outside community, and too slow to adopt modern methods. The biggest problem he found was the low number of workers who had any real training. "The employment of registered nurses in very limited numbers," he wrote, "possibly a dozen being employed at a training school of three-thousand patients, reflects the limitation of personnel available for this work."[9]

At Fernald, the problem of finding qualified help extended even to the medical staff. Most doctors did not want to work in a place where the patients were difficult, the environment was depressing, and the level of medicine being practiced was rudimentary. These obstacles meant that the two biggest sources of new doctors for Fernald were immigrants and physicians who had recently retired from state psychiatric hospitals. They were permitted to work for short terms at state schools in order to build up their pensions.

Among the psychiatrists hired at Fernald in 1957 were Carmelo D'Arrigo, an Italian Army veteran with acknowledged language difficulties, and sixty-eight-year-old Jacob Norman, who was, himself, recently released from a mental hospital. These gentlemen would be joined by a sixty-five-year-old Hungarian psychiatrist named Joseph

Bakucz, a war refugee who spoke with a heavy accent and was not yet licensed in Massachusetts.

As it turned out, Dr. Norman wasn't ready for Fernald. He soon developed what Superintendent Farrell called "bad personality problems" and was dismissed. Drs. D'Arrigo and Bakucz soldiered on. Often, they were called to interview runaways who had been returned by police. Time and again, they would sign orders placing these boys in Ward 22. On some occasions the population of the prison ward climbed as high as 50 young men. They were guarded by a staff of just eight, which meant that on any single shift, it would be rare to find more than two on duty.[10]

Whether they were in 22 or the BH, the State Boys asked endless questions about why they were being held at Fernald and how, or when, they might get out. Most attendants changed the subject or answered with questions of their own.

"Where would you go if we let you out?"

"Are you smart enough to take care of yourself?"

"You think you know better than Dr. Farrell?"

Every once in a while, a kindly attendant would halfway agree that a boy like Fred Boyce didn't belong at Fernald. One of these sympathetic staff members was an older woman named Regina Shaw. She would sometimes say that she understood Fred's complaints, but she always caught herself before she admitted that he didn't belong in state custody. Her supervisor, Patrick O'Callaghan, also found these questions difficult. He would say something about the general unfairness of life and change the subject.

"Of course a boy would try to run away," recalled O'Callaghan at the age of eighty-eight. "He's lonely, he's getting nothing out of the place, so he runs. He's caught and what do they do? They take him to 22, strip him down to his shorts, shave his head, and lock him up. He might be there weeks, with older, tougher boys, and when he comes back, he's tougher, too."

A tall, slender man with blue eyes, blond hair, and an Irish accent, O'Callaghan wore a crisp white shirt, a narrow tie, and pressed dress pants every day that he worked at Fernald. He had a strong religious faith, felt a Christian duty to the boys, and tried to make their life in the BH more normal. He installed full-size mirrors in the bathrooms—until then there had been only tiny ones—so they could see to shave and comb their hair. He allowed the boys to smoke at certain times in certain parts of the building. He supplied newspapers and magazines to the day rooms and was surprised by how many boys spent hours poring over them. He once tested a boy he saw holding a copy of *Time* magazine and heard a long description of life in Egypt and Gamal Abdel Nasser's rise to power.

O'Callaghan was not alone in his effort to calm the tensions among the boys and foster some sort of normalcy. An attendant named Mr. Settipane became a favorite among the boys because he kept order in a calm way, and while he threatened punishments, he rarely followed through. He also had a peculiar way of talking. He said everything twice. If things got loud in a ward he would say, "I'll punish you! I'll punish you!" Once they caught on to this quirk, the boys in the BH would tease him by answering in a similar way—"We'll be good! We'll be good!"

Like O'Callaghan, teacher Kenneth Bilodeau often took up a student's cause, especially if it meant that he or she would get a moment of relief from institutional life. In one case, he heard that a sixteen-year-old living at the Girls Home had been invited to an off-campus dance by a boy she had met during a visit to her family home. When a psychiatrist denied her permission, Bilodeau ignored him. He helped the young woman get a dress, and on the evening of her date he walked her to the Fernald gate to meet the young man. Bilodeau was there again when the two returned.

"I met the boy and the girl as they came back and watched them as he walked her up to the Girls Home," he would recall. "All the other girls were lined up at the windows watching. When he kissed her good night, they started yelling and cheering."[11]

Beyond the favors that came from Bilodeau, O'Callaghan, and a few others, the State Boys found some hope in the arrival of Abigail Bacon, a social worker who began a new program to train older boys to be paroled to work and live at Emerson Hospital in the town of Concord. Bacon visited the schoolhouse and told the boys that the brightest and best behaved would be chosen for this experiment, which they called the Program. There wasn't much to this program. Boys were to be given some counseling. They were then going to get jobs and supervised housing on the outside. The most important thing about this program was the hope it offered, and every boy in the BH began to imagine he would be the first chosen. Weeks later, the entire BH was surprised to hear that quiet little Albert Gagne, who many considered to be of only average intelligence for Fernald, would go first.

Considering matters from Abigail Bacon's point of view, it's easy to see why she called Albert's name first. No Fernald resident had shown himself to be more reliable. Throughout Albert's record were notes indicating he was ambitious, eager, and agreeable. In 1953, Mildred Brazier called him a "trustworthy errand boy." Two years later, it was noted that he was a "good worker" at Dr. Benda's laboratory. Albert had been so dependable in the lab that he had been given a white coat and assigned to assist pathologists with autopsies. He was not put off by the blood, or by handling the human tissue and whole organs he was asked to place in jars filled with formalin. He was, in a way, unshakable, which made school officials optimistic about his ability to adapt to the outside.

Many State Boys felt disappointed when they learned that Albert Gagne had been picked before them. Knowing that there was a door that opened to the outside made them anxious to use it. The fact that a meek boy like Albert had been allowed out first made them think that only a goody-goody type, one who lacked the courage to rebel, had a chance.

Any teacher, social worker, or attendant with access to the State Boys' records would have understood why they were so agitated. A good number of the boys in the BH truly believed they were more normal than retarded, and they were right. Fred Boyce's record noted, "He is certainly not feebleminded." And in Joey Almeida's file was the phrase "no real evidence of this boy being significantly retarded."[12]

All through the summer of 1957, the residents of the Boys Home talked about getting out and tried to get free of the place. Some were too cowed by years of captivity to actually try. The others, those who did try to escape, called these timid ones "chickenshit" and sneered that they would die in Fernald.

In all, twenty-seven escapes were reported in the three months of summer. One boy, Richard Williams, got as far as Jackman, Maine, near the Canadian border, before he was caught by police and sent back to Ward 22. (Williams was the one who had tried to kill himself by drinking green paint.) In August, Charlie Hatch escaped and lived on the streets for almost two weeks before the police in South Boston caught him. When he returned, he complained bitterly to the other boys, telling them they let the attendants treat them like animals and that "something has to be done to stop all this."[13]

While some State Boys sought freedom on their own, many of the Home Boys, whose families visited on Company Sunday, begged for them to intervene to get them out. These pleas were often made in the day room, as families prepared to depart. Cries of "Get me out of here" were typically matched with the promise "I'll be good."

Although the begging usually did no good, every once in a while a family member listened and acted. In the fall of 1957, it happened for Joe Zupokfska—pronounced Zuh-*pof*-ska—a teenager from Dracut whose parents had committed him to Fernald. Zupokfska was a muscular boy who combed his greased black hair straight back and wore T-shirts and jeans with the cuffs rolled up. He had been old enough to go straight to the BH when he was admitted. In his first week there, Fred Boyce had befriended him and helped him adapt.

"Joe was one of these kids who came to Fernald not really knowing how to fight," recalled Fred. "He was tough enough, really, but he didn't have experience with defending himself. At the BH there were thirty-six kids who were going to test him out, initiate him. I told him that the next time someone came to pick on him, all he had to do was pop him a real good one. Everyone would see what he did, and learn from that."

Zupokfska got the chance to try Fred's advice when he had a run-in with Charles Dyer, a sixteen-year-old who had been at Fernald for just two years. Dyer had good reason to be in a fighting mood. He had just endured a rare visit from his mother, and it had gone terribly. She had promised to get a lawyer and petition for his release, but only if he agreed to live with her, find a full-time job, and turn over most of what he earned to her. Long furious over the fact that she had abandoned him to Fernald, Charles had told her he would rather stay locked up than be free under those conditions. Moments later, he picked a fight with Zupokfska. The newcomer responded in an unexpected way. He hit Dyer in the mouth so hard that his tooth came through his lip, leaving a large, bleeding wound.

In his fight with Zupokfska, Dyer suffered a little pain but also got rid of some of the anger he was carrying. But while he became a bit more relaxed afterward, the battle seemed to have the opposite effect on Zupokfska. He became willing, even eager, to fight with other boys. He got so cocky that he challenged Fred, who wrestled him to the floor of the ward and demanded a cry of "uncle" before letting him up.

Although other boys stopped bothering him once he fought back, Zupokfska hated life in the BH. Since he had attended public school, he could see that classes at Fernald were a waste of time. He told his mother about this at every chance he got, and eventually she applied for his release. Before he departed, Zupokfska thanked Fred Boyce for his help and gave him the one thing of value he possessed, his address on the outside.

"If you ever get out," said Joe Zupokfska, "come see me first." He pressed a piece of paper into his hand; on it was written "25 Varnum Road, Dracut."

The departure of one boy made no difference in conditions at the BH, because whenever one inmate left a ward, it seemed two were moved in. The space between the beds got smaller and smaller, and privacy disappeared entirely. The same overcrowding happened at all the buildings, including Ward 22. Even the day rooms, which were supposed to relieve the congestion in the wards, became more crowded. It was hard to get a moment of quiet to read the newspaper, or to watch television. But when they could focus on the images on the front page or the nightly news, the State Boys caught snatches of truly exciting events and ideas.

In the fall of 1957, the press was filled with reports on the battle over the integration of public schools in the South. In September, armed guards escorted nine "Negro" students into Central High School in Little Rock, Arkansas. In early October, scores of white students walked out in protest. Later, 6,000 whites and blacks gathered for a day of prayer for peace. The drama and turmoil in Little Rock was covered nightly on the news. John Chancellor of NBC became famous because of his reports, which, for the first time, put the drama of the civil rights movement in full view.

The State Boys viewed these events on TV, and many understood that they were watching an oppressed minority assert its rights. Though it would be safe to assume that some had racist feelings—after all, they called Fred Boyce "nigger boy"—they also related to the underdog. And it was clear to them from the coverage on TV that those who struggled for equality could find support and even sympathy when their case was made well.

Decades later, it would be impossible to say how much their awareness of the civil rights movement affected what some of the State Boys did on November 4, 1957. It's likely that, along with the rest of the country, they were just as interested in the USSR's dog-bearing spacecraft, which had been launched on Sunday, November 3, and sent the nation into a Cold War dither.

But Joey Almeida would become convinced that what occurred on that fateful day in Ward 22 was inspired, at least in part, by the bravery of those teenagers in Little Rock. Others noted the theme of teen rebellion in pop culture and how movies and even song lyrics inspired defiance. Add the daily humiliation, frustration, and rage that came with life in Fernald, and an uprising became inevitable. "We just had that feeling that something had to be done, to save us and to bring down that place," Charles Hatch would say decades later.[14]

Joey had started the day with breakfast in the dining room on the ground floor of Ward 22. He had then taken a short walk to the Manual Training Building. It was cold outside, below 30 degrees Fahrenheit, and the sun shone through broken clouds. Red, orange, and yellow autumn leaves littered the walkways. A stocky boy with thick curly hair, Joey walked quickly, covering the few hundred yards to the Shop Building in just a few minutes.

Once he arrived at the Manual Training Building, Joey reported to the shoe shop. Most of the work done there involved refurbishing old boots, so he set to work with a heavy needle, using waxy cord to join new soles with old uppers to make serviceable pairs. At noon, Joey went back to the dining room at 22 for lunch, and then to the day room to relax before afternoon classes at school.

Forty boys were packed into the day room. Some watched TV. Others played cards or read the newspaper. The front page of the morning *Boston Globe* screamed DOG ALIVE IN SPACE. The paper explained that the Russian dog was in orbit 1,000 miles above the earth and that its craft, *Sputnik II,* would be visible in the sky over Boston that night. The region's Catholic Archbishop Richard Cushing (soon to be a cardinal) bemoaned the apathy Americans showed toward "the greatest enemy [communism] our civilization ever faced." The Soviets said the dog, a female Laika named Curly, was in good condition after twelve laps around the planet.

Curly was a minor curiosity for the boys in Ward 22. They were focused instead on a plan that Charlie Hatch had devised in the time that Joey Almeida had been at work in the shoe shop. Fed up with the treatment he had received at Fernald, Hatch had told the others that the time had come to assert themselves. They were nearly men, he pointed out, big enough, strong enough, and certainly angry enough to take on the guards and confront Fernald's administrators.

Age and physical size were two important factors. But the young men in Ward 22 had also been primed for action by their increasing contacts with the *outside.* Boys who had escaped and been captured told stories of how they survived on their own, giving others courage. At the same time, the increased number of visitors to the school—Bea Katz and others—had given many Fernald residents a chance to practice conversation and discover they were not as deficient as they feared. Finally, the younger teachers and counselors hired at the school had encouraged the young men of Fernald to think of themselves as capable and competent. With so many factors preparing the young men of 22 to act, Hatch didn't have to push very hard to get a dozen of them to volunteer to drive out the attendants and take over the building. (It also didn't hurt that Hatch, with his black slick-backed hair and powerful build, looked the part of a teen rebel.)

Joey, who had held fast to his belief that he never belonged at

Fernald in the first place, volunteered to be part of the fight. At about three o'clock in the afternoon, twenty-five boys who were too frightened to join the uprising were gathered in the day room so they could leave quickly when the action began. Then one of Hatch's recruits went to a clothes closet and started a small fire. An attendant ran to pull an alarm. As he did this, a State Boy named Bobby O'Brien slipped behind him and seized him by his belt and shirt collar. With a rush, O'Brien ran the attendant to the door and pushed him outside. Behind him, a crowd of boys began yelling and cheering like a mob watching a bar fight.

When the second man on duty came downstairs from the ward to check on the fire alarm, he was told to get out or he would be thrown out. He accepted the offer, and was followed out the door by the group of boys who had chosen to leave rather than fight. Of the fifteen who were left behind, several went to put out the fire while others locked the door and jammed the keyhole with buttons they tore from their shirts. Charlie Hatch went into the ground-floor attendants' office, snatched a key that was hanging from a hook on the wall, and went upstairs to unlock the cells used for solitary confinement.

Some of the boys who occupied the isolation cells hadn't waited for Hatch to come let them out. Charlie Dyer, his lip now healed from the fight with Joe Zupokfska, had previously hidden a wire coat hanger under his mattress. He took it out and bent it into a long hook. In his cell, the window in the door was just an opening—no glass—so he snaked the hanger through it and began feeling for the old-fashioned bar-style latch. He caught it, pulled hard, and then pushed on the door. When it opened, he rushed out into the hall and released the others from their cells.

"We were really fed up," recalled Dyer many years later. "We wanted to do anything to show them they couldn't do this to us anymore. I told them to pile up the mattresses in my cell. I got a bunch of newspapers and set the whole thing on fire. We broke the windows to let the smoke out and it just burned. The room didn't catch fire, but the mattresses sure made a lot of smoke."

Down on the first floor, Joey and some of the others turned their attention to defending the ward from anyone who might come to put down their insurrection. Here their intimate knowledge of the mechanics of the institution—especially how security and fire systems worked—gave them an advantage. They unrolled a fire hose, which was connected to an inside hydrant, and smashed a small window beside the doorway. They then took turns standing guard, ready to turn the harsh stream from the hose on anyone who approached.

While defenses were being established on the first floor, the boys on the second level of the building went a little berserk. Charles Hatch led a small group that piled newspapers onto dinner trays and then climbed up to place them in the rafters. They then set the papers on fire, intending to burn the wooden beams. The fires on the trays kept going out before the rafters caught the blaze. Frustrated, Hatch and the others then went into the bathrooms and tore sinks and toilets loose from their fittings. They used broken pipes to smash mirrors and then hurled the porcelain fixtures against the tile walls. Each one made a satisfying noise as it broke.

As the boys attacked the interior of the building, teachers and attendants joined Malcolm Farrell on the side of the hill that overlooked Ward 22. Word of the riot had spread quickly across the campus. Fred Boyce and other State Boys who were not in Ward 22 at the time were prevented from going to the scene, but attendants told them what was happening. Sirens could be heard in the distance as trucks from the Waltham Fire Department sped to the school. By the time the firefighters arrived, the smoke had stopped coming from the broken windows. When Superintendent Farrell used a megaphone to ask if the rioters were all right, the boys hollered back to say they had put out the blaze and that no one had been injured by either the fire or the rioting.

It's possible, decades later, to sympathize with Farrell as he faced the rioters inside Ward 22. With 2,600 inmates, inadequate staff, and mounting pressure from the outside world, he was dealing with greater burdens than anyone who had ever been in charge of Fernald had faced. He knew that many of the angry young men of 22 didn't belong at Fernald. He preferred to pacify them rather than confront them. But now they were challenging him for physical control of the institution itself. Something had to be done.

Local police had followed the fire department to Fernald, and Farrell turned first to them. He wanted the rioters rooted out, but he would not allow the use of brute force, or even tear gas. A handful of officers agreed to try to retake the building with the help of some attendants. Armed with pry bars supplied by the maintenance department, they marched down the hillside to the heavy steel door of the building.

Inside the ward, Joey and Charlie Hatch ducked low and peered out the broken window beside the door. They watched the police and the attendants creep toward the building. When they were just feet away, Joey hollered for Charlie to turn the valve on the hydrant. As the water surged down the line, Joey raised himself up and pointed the nozzle out the window.

The force of the water made the nozzle hard to control, but Joey managed to blast the men outside. For a moment, they stood fast, shielding themselves with their arms and leaning against the water. Then Joey was surprised, and the men were sent running, when the water turned from cold to hot. Apparently, the fire-fighting system had been designed so that in the event of the loss of the cold-water supply, which had been turned off, the fire lines would be charged with water from Fernald's central boiler.

On the hillside, dozens of people gathered to watch what would happen next. Pat O'Callaghan was just arriving when someone screamed from an upper-floor window. "Back off or we'll burn the fuckin' place down!"

Nothing like this had ever happened at Fernald, and O'Callaghan found Malcolm Farrell struggling to decide his next move. He didn't want to use force, but he feared it might be necessary. State police officers arrived from the nearby Framingham barracks, but when they advised him to use tear gas, he chose to wait. The boys were bound to get hungry, and tired. It would soon be dark. As long as the rebels refrained from setting any new fires, time was on the superintendent's side.

Feeling as much trapped as triumphant, many of the boys inside the building knew they would eventually give up, but they wanted to exploit their position—mainly the freedom to vent their anger—in every way possible. A large contingent went into the ground floor office, pulled open file drawers, and began dumping out case records. They tore up some of them, and burned others in a wastebasket. A few recognized their own names on the files, but they didn't take time to read them carefully. Most couldn't read well enough anyway.

Once the files were scattered, the boys went to the kitchen, opened the refrigerators, and feasted on the hot dogs and bread that they found inside. After gorging themselves, they became worried about having destroyed all the toilets in the building. When some boys began using the tile floor as a urinal, the three leaders of the coup—Joey, Charlie, and Bobby—began to talk about what they should do next.

At dark, emergency floodlights were brought to the roofs of nearby buildings and switched on, casting the whole area in a shadowy half-light. A bus full of state police officers arrived to reinforce the police who were already on the scene. They stood in the gloom, growing colder and more impatient as the hours passed. Inside, some of the boys felt a sense of foreboding. This standoff couldn't last forever. Others were too filled with adrenaline and the excitement of asserting their power to think ahead to the ultimate end of the situation.

By 10 P.M., seven hours into the siege, Mental Health Commissioner Jack Ewalt arrived wearing what appeared to be his bedroom slippers. A psychiatrist with a national reputation in the field of hospital services, Ewalt had a general approach to troubled young men that was fairly compassionate. In a previous brush with a notorious case, his report on an eighteen-year-old murderer had helped persuade the governor of Massachusetts to spare the convict's life. With the Fernald rioters, Ewalt was again reluctant to suggest a violent approach, even when some of the boys tried to bait him.

It happened when Ewalt moved down the hillside, with Pat O'Callaghan beside him, and hollered up at the second-floor windows. He tried to persuade the rioters to give up, and promised they would be treated fairly. O'Callaghan never forgot the response shouted from above.

"Fuck those bastards," he heard one boy yell to another. "We should just burn this fuckin' place down!"

Ewalt and O'Callaghan retreated. In the next hour, the commissioner devised a plan. He told state and local police to take up positions surrounding the building, making sure that the boys inside could see they were posted in substantial numbers and armed with pistols and rifles. They were under orders to use no force, but the boys wouldn't know it.

Just before 11:30, Commissioner Ewalt again padded up to the door of Ward 22 and shouted for the boys to give themselves up. They had good reason to give in. First, they were out of food. Second, with the windows all broken, the building was getting cold. Finally, the stench from the bathroom was becoming unbearable.

Inside, the group turned to Charlie Hatch and Joey Almeida to make a decision. The two leaders figured that sooner or later they would have to face the consequences of what they had done. One of the boys went to the broken window by the door and shouted surrender.

In less than a minute, a police paddy wagon backed up the walkway and stopped within a few feet of the entrance to the building. Because he knew every boy in the ward, Kenneth Bilodeau was sent down from the hillside to call out their names as they emerged. One by one, the boys came out. Their names were recorded by a police officer, and then they stepped up into the truck.

In just minutes, fifteen boys were seated on the benches in the truck and the doors were slammed shut. No one told them where they were going as the vehicle lurched forward, jostling them against each other. They asked the driver where they were headed, but he wouldn't respond. They took turns peering out the tiny window in the back of the truck, trying to recognize landmarks. After a while, they gave up on this and sat back in the darkness and the silence. Joey listened to the truck's motor whine and felt the thunk each time the driver shifted gears. For the first time in years, he did not feel angry. He felt scared.

When the paddy wagon stopped and the doors were opened, flood-lights almost blinded the boys inside. A gruff voice ordered them to get up and out of the truck. As the boys obeyed, their eyes adjusted and they could see that they had been brought up to the doors of another big institution. Here there were not attendants, but uniformed guards, and the toughest one of all was screaming at them.

"Who's the really tough guy?" he demanded to know. "C'mon boys. Who's gonna show me how tough he is? This ain't no fuckin' school you're in now."

They were in the Bridgewater State Hospital for the Criminally Insane. Though technically a psychiatric facility, Bridgewater was more like a maximum-security prison, with iron bars, a big security fence, and a population of potentially dangerous adult inmates. In 1957, Sacco and Vanzetti, the controversial anarchists of the 1920s, were probably the most famous of Bridgewater's alumni. Both had been held there for evaluations, which had determined that they were sufficiently sane to be tried for murder and then executed.

In their first moments at Bridgewater, the State Boys were herded into a nearly barren room the size of a basketball court. Following orders, they stripped and then bent over to be searched by a guard wearing rubber gloves. Then, still naked, they were ordered to stand on a section of the floor that was marked off by a painted yellow line. They were then summoned, one by one, to sit on a metal chair beside a desk that was manned by a ferocious-looking guard who clenched a wet stogie between his lips. It went something like:

"Name?

"Age?

"Hometown?

"Do you know where you are?"

After his interview, each boy was ordered to shower and then dress in prison clothes—a denim shirt, a pair of dungarees, and black shoes. Each was assigned to an individual cell, where the furnishings included a mattress and a slop bucket. For the next few days, the rebels of Ward 22 were confined to these cells except during mealtimes, when they went to a dining hall. From time to time, they were asked the same questions about their names, ages, hometowns, and whether they knew where they were. Sometimes they would also have to tell a guard, or a prison doctor, the name of the president of the United States, and the day, month, and year.[15]

Because outside police and firefighters were called onto the Fernald campus, the uprising at 22 could not be kept secret. (In contrast, few if any escapes, even those involving large groups, were ever made public.) Newspapers in Boston and Waltham gave prominent play to the event, even publishing photos of the ruined interior of the ward. However, as a sign of the times, the boys who were carried off to Bridgewater were described erroneously as "delinquents," a designation never applied inside the institution. At the time, much of America was deeply concerned about the supposed menace of juvenile delinquency. Congress had investigated the problem, and acres of newsprint had been devoted to exposés on the subject. It was, therefore, convenient and easier to call the Fernald boys delinquents—readers understood the label—rather than deal with the complexities of their actual status as abandoned, neglected, or learning-disabled children.[16]

The staff at Fernald surely understood at least some of the circumstances that had brought the boys who rioted to the institution in the first place. While a few may have come into state custody as true criminals, the vast majority were long-term wards of the state with dubious diagnoses. They were not delinquent. In Waltham and at similar state schools around the country, a growing number of professionals were coming to recognize the absurdity of holding hundreds of relatively normal children and allowing them to become angry, undereducated, desperate adolescents. After the riot, the danger in this practice could no longer be denied. And in retrospect, Fernald would never again be the same. Never again would a prison ward be used to hold large numbers of teenage boys. Never again would young men like Joey Almeida and Charles Hatch accept that there was nothing they could do to affect their own fates.

Indeed, the changes that took place inside the State Boys themselves were more profound than the changes the riot brought at the institution. The rebellion had been deeply satisfying in many respects. The intelligence, organization, strength, and courage required to pull it off could not be denied. Very quickly, the uprising took on a mythic quality, a story that affirmed their sense that they were no less competent than the staff who were supposed to mind them. In their defiance, the boys who rebelled had defined themselves as men worthy of freedom.

Two days after their incarceration at Bridgewater, the fifteen State Boys were brought to a courtroom inside the prison walls. There, during a brief hearing, Fernald Superintendent Farrell told a state judge what had occurred at Ward 22 on a night when the world's attention was turned skyward, toward *Sputnik II*. Farrell said that many of the rioters were too unruly to be returned to Fernald, but allowed that he would welcome a few of them back, eventually. He mentioned one by name—Joey Almeida.

When Farrell finished his testimony, the judge asked him a few questions and then divided the rioters into two groups. Eight boys were formally imprisoned at Bridgewater for an indefinite stay. The remaining seven were assigned to the prison hospital as temporary inmates whose cases would be reviewed in thirty-five days.[17]

Though Joey Almeida was in this second group and was told he would eventually return to Fernald, no one said exactly how long it would take. He would be served Thanksgiving dinner on a tray at the prison cafeteria and greet Christmas morning in his cell. (Like the others, his one gift would be a package from Archbishop Cushing that contained a toothbrush, toothpaste, and a bar of soap.) As New Year's Day came and went, Richard Williams and Joey talked over their prospects. Richard guessed that he would be returned sooner, so he promised to "take care" of the Fernald girl that Joey considered his girl-friend.

Time passed more slowly at Bridgewater than it had at the state school. Though they occasionally saw a psychologist or psychiatrist, the boys received no other services, no schooling, and no vocational training. They ate their meals at different times than the other inmates, so that they wouldn't be harmed by them. They were allowed to use a small exercise yard. At night, the boys were confined to one-man cells. Though they were separated from the older inmates, they could hear them shouting and screaming through the night.[18]

Back at Fernald, Malcolm Farrell reported to his trustees that what had occurred in the takeover of Ward 22 was a riot led by a handful of trouble-makers. He reminded them that only 15 of a possible 2,600 inmates at the school were involved. Though Farrell's report suggested the riot was just a spontaneous, isolated event, anyone curious about the cause could have found a big clue in the last sentence of the *Boston Herald*'s article on the uprising, which said, "Dr. Ewalt said last night there have been reports of grievances from the 15 during the past three weeks."

At a meeting of all the state school superintendents from around the commonwealth, it was also acknowledged that the Fernald riot was the result of a long buildup of tension. According to the minutes of that meeting, "The superintendents all felt that they had similar cases in their institutions," and that they were "sitting on a powder keg." This document also confirmed what so many of the State Boys believed. ". . . the institution," it concluded, "is not geared for their care and treatment."[19]

Confined to the Boys Home during the fracas at Ward 22, Fred Boyce had gone to sleep on the night of November 4 not knowing the outcome of the standoff. When morning came and the shift change brought new attendants, the news began to spread. The boys at 22 had overwhelmed the attendants, trashed the building, and then given up. Now they were in Bridgewater. Fred worried most about the smallest, youngest boy in the group, his friend Joey.

For years, Bridgewater had loomed over the State Boys like a monster. Attendants would threaten to send them there and add to the terror by describing the terrible things that would happen there. "They'll put you in with the crazies. They'll put you in a straightjacket. They'll zap you until you're a zombie." According to the rumors that flew around Fernald, the fifteen boys at Bridgewater had been subjected to beatings, shock treatments, straightjackets, and lobotomies. In fact, the worst they suffered was isolation, loneliness, and confusion. They were kept separate from the adult inmates and, as long as they were quiet, received reasonable treatment from the guards.

One of those who refused to be quiet, Charlie Hatch, did suffer some consequences. After using a chair to break windows, lights, and the plumbing fixtures in his room, Hatch was set upon by five guards, some of whom hit him with blackjacks. He was then wrestled into a higher-security area and injected with enough sedating drugs to knock him unconscious.

At the Boys Home, Fred continued to fret about Joey and the others. It seemed to him that something very basic had changed. The state authorities were now willing to make good on their most serious threats. This realization made him feel even more desperate to get out.

He had learned that he needed money and, most importantly, a place to go where he could count on real help. A place like the Zupokfska house on Varnum Road in Dracut.

On a Saturday, after he had looked at the morning newspapers, Fred asked one of the attendants at the BH if he knew where Dracut might be.

"Why?" came the response.

"I read something in the paper about a fire there. I never heard of it before."

"It's up north by Lowell, on the way to New Hampshire."

Days later, Fred would ask Pat O'Callaghan about the city of Lowell, where it was, and what kind of work was available there. He would pretend to care as O'Callaghan talked about the mills that produced shoes, textiles, and plastics. Eventually, Fred got him to say that state Route 3 led from Boston to Lowell. When O'Callaghan asked why Fred needed to know, he said, "No reason. I was just wondering."

Armed with the basic knowledge that might get a hitchhiker to Dracut, Fred then set to gathering money. For this, he would depend on the more retarded of the Home Boys. On Company Sundays before their visitors arrived, he told them to save any money they might receive from their relatives. "You bring it to me," he told them, "and I'll trade mine for one of yours."

On the first Sunday in December, Fred collected more than $5 by trading pennies for quarters, dimes, and nickels. Robert LeBrun, an easygoing Home Boy who cherished his friendship with Fred, proudly handed over four quarters. He seemed so pleased to help that Freddie decided to push the game a little further.

"LeBrun," he said. "Do you know what paper money is? Next time, you get some of that."

That night, after the ward had gone to sleep, Fred shimmied up a water pipe in the lavatory and found an open space in the ceiling where the pipe passed through to the floor above. He tucked a little bag filled with quarters into the space. As he turned to slide down, a watchman just happened to come into the room. Fred, desperate to conceal what he was doing, began scratching himself and howling like a monkey.

"Hoo, hoo, hoo!"

He then slid the rest of the way down and, when he reached the bottom, pretended to be awakening from a dream. He would never know if the act was convincing or if the attendant just didn't care. Either way, the result was what Fred wanted. The watchman told him to go back to bed, and the stash at the top of the pipe remained undiscovered.

A month passed before the first Sunday of January, when the Home Boys went on their visits and dutifully returned to trade quarters for pennies. LeBrun arrived with a big grin on his face and a pocket bulging with cash. Fred hurried him into the bathroom and, when all was clear, said, "C'mon LeBrun, what have you got?"

"Paper money!" said the other boy as he excitedly searched his pocket and pulled out a fistful of bills. He opened his hand, to reveal fives, tens, and twenties—all of it from his family's Monopoly game.

Fred's disappointment was real, but he was more amused than upset. He had already taken in several more dollars in silver and knew he had enough to buy food for a day or two on the road. He said thank you and took the pastel bits of paper from a proud and happy LeBrun. In the weeks that followed, Fred called him by a new nickname—Monopoly.[20]

SEVEN

After lights-out, Fred Boyce lay awake as the other boys in his ward fell asleep. He kicked off his blanket so that the chill in the underheated Boys Home would keep him alert. He passed the time by watching the clock and reviewing his escape plan over and over in his mind.

At about 4:30 A.M., Fred slipped out of his bed, fluffed his pillow, and covered it with his blanket. He crept to the bathroom, used the toilet, but decided not to risk the noise of flushing. He then shimmied up the water pipe, retrieved his sack of coins, and then slid down to the floor.

With the coins in hand, Fred went to the big clothes closet, where he shed his nightshirt and dressed for the cold weather outside. He then sneaked into the hallway and down the stairs to a storeroom on the first floor. There he opened a window, climbed onto the sill, and swung out onto an iron railing that guarded a catch basin near the side of the building. After he steadied himself on the rail, he was able to hop to the ground, which was covered with about six inches of snow.

Each step that Fred took produced a crunching sound as his feet broke through the frozen snow. He moved slowly down the steep hillside to Trapelo Road, stopping every few feet to look around for security officers. Once on the street, the footing was better, and he walked quickly to Waverley Square. There he found shelter in the doorway of a coffee shop that was not yet open and waited for the traffic to build so he could hitch a ride.

Fred had learned in his previous escapes that the way that he talked—the way all State Boys talked—was different from how people spoke on the outside. They had trouble using expressions. Instead of saying they would "brush up" on a subject, they might say they would "brush on" it. Their other malapropisms included "catfooting" instead of "pussyfooting" and "doping around" instead of messing around. Inside Fernald, they could speak this way and be understood, so they were rarely corrected. Outside the institution, these habits of speech made them stand out.

State Boys were also blunt, quick to tell someone to "fuck off" or "get out of my face." And they got so nervous when they were asked questions that their answers could seem odd or stupid. If a motorist asked, "Do you know someone in Dracut?" a flustered State Boy might blurt, "I don't know." For this reason, Fred thought carefully about how to handle questions. He decided to keep his answers short, and to make it clear that he knew where he was going and he had good reasons for the trip.

"Dracut" was all he said to the first driver who stopped for him. The man stared for a moment and then told him to stand on the other side of the road because he was headed south and Dracut was to the north.

After he crossed the street, Fred stuck out his thumb again and soon had his first ride. He traveled back roads through the towns of Concord and Carlisle. There he got a ride from a man who was headed to Lowell, which was next to his destination. This driver quizzed him thoroughly.

"You got family there?"

"No."

"You work up there?"

"Nope."

"Then why are you going?"

"To see a friend."

Soon the car pulled to the side of the road—by this time Fred was on Route 3 in Wilmington—and the driver told him to get out. Fred obeyed, closed the door, and the car sped away.

Fortunately, Fred was close enough to his goal that one more ride brought him to an intersection on Varnum Road. He thanked the driver and then had the good luck to walk in the right direction. A little more than a mile down the road, he found No. 125, a plain, two-story, wood-frame house with overgrown shrubs and snow piled on each side of the door.

As Fred turned up the driveway, Joe Zupokfska just happened to come out the door. He was surprised but pleased to see his old friend from the BH. He brought Fred inside and introduced him to his mother. (Joe's father no longer lived with the family.) Mrs. Zupokfska had believed her own son's reports of life at Fernald and was sympathetic to Fred. She agreed to let him stay, at least temporarily. He would sleep downstairs, on a sofa.

On the afternoon of January 12, 1958, the two young men, who now considered themselves *former* State Boys, drank sodas, listened to the radio, and made a plan to look for jobs. Joe knew a house painter who was working in the neighborhood and needed a helper. They agreed that the job would be perfect for Fred, in the short term. They thought that eventually they would each find work in one of the local mills. And perhaps, in a few weeks or months, they would get their own apartment.

After a day or two passed, Fred found the painter and got the job. It involved fetching supplies, cleaning up after his boss, and helping coat ceilings with white paint. At the end of a week, he was paid in cash and handed most of it over to Mrs. Zupokfska. On the weekend, Fred and Joe walked to downtown Lowell, where they hung out in a soda shop and listened to a jukebox. It was the most fun Fred had had since his vacation with Mary Mone and her husband.

The only hitch in Fred's adjustment to the Zupokfska home involved Joe's sister. At first, she showed absolutely no interest in him. Then, after a week had gone by, she complained that she felt uncomfortable with a stranger sleeping in the living room, which she passed through on her way to a downstairs bathroom. Fred was moved into Joe's room, where they shared a large bed, and the problem seemed solved.

Fred lived what he thought was a normal life for eight days. On the ninth day, he went to work painting and returned to the house on Varnum Road just before five o'clock. He found Mrs. Zupokfska sitting alone at the kitchen table. Joe and his sister were not home. She offered him a soda, which he took out of the refrigerator, and she told him to sit down.

In halting sentences, Joe's mother told Fred that the arrangement they had made wasn't working out.

Interrupting, Fred told her that he could find a better job at a factory in Lowell and pay her more.

It wasn't a matter of Fred's paying more money. Mrs. Dorothy Zupokfska explained that she was dependent on welfare, which she would lose if authorities discovered that Fred was living with her and giving her any money at all. She couldn't take the risk of losing her check or, worse, being charged with breaking whatever laws came into play when an adult harbored a runaway ward of the state.

Fighting tears, Fred asked her if she understood how bad Fernald could be, and why he had fled in the first place. He sputtered something about running away again. And then he asked where Joe and his sister were. Mrs. Zupokfska looked at him with a pained expression and told him it was too late for him to run. The police were going to arrive in a matter of minutes. "They said they would take good care of you, and make sure nothing bad happened."

Two Dracut police officers—their last names were Dacey and Wardman—knocked on the door while Dorothy Zupokfska was still talking. When they came in, they tried to reassure Fred. He begged them to leave him alone. They said they understood why he was upset but had no choice in the matter. When he said that the attendants at Fernald would beat him, they promised "we won't let anyone do anything to you."

Gradually, Dacey and Wardman talked Fred out the door and into their patrol car. They brought him first to the local police station, where they filled out some paperwork and entered his capture in the daily log. They also contacted Fernald and received detailed instructions on how to handle the transportation and delivery of a runaway "moron." By the time they were finished and took a dinner break, it was approaching 10 P.M. They fed their prisoner a sandwich and then loaded him in the back seat of their car for the long drive to Waltham.

With Ward 22 in a postriot shambles, the staff at Fernald had begun using isolation rooms in the North Building as a substitute. The boys at the BH referred to the place as the Snake Pit and considered it a symbol of degradation and humiliation.

Everything he had ever felt about the North Building and its inhabitants flooded Fred's mind as the patrol car stopped at the door. He argued with Dacey and Wardman, but they told him they were at the right place. Fred stayed quiet until they had brought him inside and began looking for the office. That's when he broke for the door.

While Wardman hollered for help, Dacey chased Fred down the hallway, grabbing him by the collar of his shirt just as he touched the doorknob. Two attendants came out of the first-floor office and followed Wardman toward Dacey and Fred. Though they were four men against one teenage boy, the fight went on for a minute or more. Fred delivered a punch to one of the attendants that split his lip. He kicked Officer Dacey in the shin. Finally, with each man seizing one of Fred's limbs, they managed to subdue him and haul him up to the second floor.

Upstairs, Fred was dragged down a hallway that looked like the cellblock off the common room in Ward 22. The men who had him by his arms and legs stopped in front of one of the doors. While three of them pinned him on the floor, the fourth turned the knob of a lock and opened one of the cells. The men still holding Fred started to strip off his clothes—another humiliation—but he fought so fiercely that they gave up after removing just his shoes and his pants. They shoved him into the cell, and slammed the door. Fred screamed for them to let him out. He was told to sleep, that he would get a chance to say his piece in the morning.

Inside the little room, which was lit by a single bulb in a ceiling fixture, Fred found nothing but a small bed on a metal frame. Still panting from the fight, he yanked the mattress off the frame and threw it against the door. He then tipped over the bed frame and slammed it against the wall. Exhausted, he slumped down onto the floor panting. He stared at the ceiling, and then at the door with its little window of glass and wire mesh, and then at the tipped-over bed. He got an idea.

Fred got up and pushed the metal frame, which was still upside down, into a corner. He then grabbed one of the legs of the bed and applied all of his weight, and all of his adrenaline-enhanced strength, against it. As he did, he felt it bend, ever so slightly. He then changed positions, so that he could bend it back. The metal gave way again. He let his hand drop to the point where the paint was creased by the bending and he felt it was warm. He had felt this kind of heat before, when he had worked copper wire back and forth to break it into various lengths for his crystal radio.

Now working with a purpose, Fred pushed and pulled on the leg of the bed frame. He got hot with the effort, so he took off his coat. As the soft steel got weaker, it bent more easily, until he was able to move it rapidly. Suddenly, the foot-long metal leg just broke off in his hands. He touched the warm jagged edge, where it had separated. It was sharp, like a chisel.

It had taken Fred about fifteen minutes to work one leg off the bed. He breathed in deeply, exhaled, and went to work on another. Back and forth. Back and forth. This time, he felt more confident, and the job went more quickly. Soon he had another hunk of steel. It was heavy, like a hammer.

With tools in hand, Fred turned to the cell door, planning to chisel a hole through it so he might reach out and open it from the outside. He saw no knob on his side, not even a plate that would cover the hole for a lock. He had to guess where to make his hole. He picked a spot about waist high and began tapping.

The North Building was a big, old brick structure with thick walls. This might explain why no one heard Fred at work. More than an hour passed as he slowly chipped away at the wood. He broke through, and then began chiseling to make the opening big enough for his hand. After a half-hour of this effort, Fred had made a hole large enough to allow him to snake his hand and arm out. He strained to reach up, and then down, but found nothing on the other side of the door. He had made the hole in the wrong spot. Exhausted, Fred slumped back on the floor, with his back to the door. He was too angry to cry out loud. He wouldn't give them the satisfaction. But as he rested, and caught his breath, tears fell from his eyes.[1]

Fred had almost fallen asleep when a sudden jolt—someone had kicked the other side of the door—roused him. It was the night watchman on his rounds. He was peering through the window and shouting for Fred to get up. He wanted to know what had happened to the door.

All the rage that Fred felt in his struggle with the bed frame and the door surged back into his body. He grabbed his hammer and chisel, quickly got to his feet, and used them to smash the window in the door. The night watchman jumped back as tiny pieces of glass fell like a handful of gravel on the floor around him. When he saw that the boy was armed, he went to a phone and called for help. Fred waited, and seethed.

In a short time, the watchman was back. Pat O'Callaghan and a nurse Fred knew as Miss Ross were with him. He liked Miss Ross, and he trusted O'Callaghan. They listened from the other side of the door as he explained what had happened, how he had found a home in Dracut, only to lose it. He said that the North Building was an insult to him, and that he would have behaved better if they had just put him in a decent place.

O'Callaghan believed him. He believed that the North Building was torture for a young man like Fred and that the institution was as much to blame for what had gone on this night as Fred. What's more, he knew it was a sign of the boy's intelligence, and the strength of his spirit, that he had tried to run away and had continued to fight for himself.

"But he had to calm down, or they never would listen to him," recalled O'Callaghan decades later. "We told him that in the morning he would be allowed to talk to Dr. Kelley [a Fernald psychiatrist] as long as he took the medication the nurse had brought for him."

Fred agreed to take the medication, insisting "it won't do anything to me." The night watchman opened the door. He and Pat O'Callaghan went in and pulled out the busted bed frame to make room for the mattress to be laid out flat on the floor. Nurse Ross told Fred to sit down. She handed him a glass of water she had brought and a single tranquilizer capsule. Fred took it and lay down on the mattress. In a few minutes, he was asleep.

In the morning, Fred awakened to the sound of someone sweeping the hallway. He peered out the window in the door and saw a severely retarded man working a broom. A metal dustpan was propped against the wall. Fred called. The man turned his head but didn't speak. When he went back to his work, Fred blurted out the only thing he could imagine might establish a connection. "Hey buddy, hulla-ma-shave," he said. When the man looked back, Fred repeated the strange word. "Hullamashave. Hullamashave." This time he got a smile in response.

Fred hoped to talk the man into bending the dustpan and forcing it through the hole he had made in the door. He thought it might make a good tool, or even a weapon. But as he whispered to the sweeper, an attendant named Mr. Bell appeared and sent the man away. He told Fred that he knew he was a good kid. He said that if he behaved, he wouldn't have to go to the Blue Room and "hose down the men" who lived in a nearby ward. Fred said he wasn't going to accept that kind of job no matter what the attendant said.

Now Bell became angry, telling Fred that he must comply with any order he might give. He was the boss, he explained, and no "reject" who was so stupid that he got himself caught after managing to run away was going to give him any back talk, let alone defy a direct order.

Though he would normally feel insulted and angered by anyone calling him stupid, Fred focused instead on taking action. There was no way he would obey this man's order to enter a ward in this building and bathe the men inside. That was the attendant's job, and Fred knew he was just trying to humiliate him. His only option, he decided, would be another escape. And to do that, he would have to get the attendant to open the door. He challenged him, saying that he "talked big" as long as he was protected by a locked door.

Provoked, the attendant opened the door. Fred let him in, and then quickly reared back to throw a punch. The attendant ducked, but instead of swinging, Fred simply dashed past him and out the door. His bare feet slapped the hallway floor as he ran away.

The fact that he wasn't wearing pants, or shoes, didn't really occur to Fred until he had descended the stairs, pushed open the door, and found himself running in snow. With his mind set on reaching his dorm and getting some clothes, he sprinted up the steep hill toward the BH, looking over his shoulder just once to see that the attendant was not in pursuit.

When he got to the BH, Fred climbed onto a porch railing, opened a window, and crawled into the day room. From there, he sneaked into a storage closet where he knew he would find clothes. As he was getting dressed, Fred decided he would go look for Regina Shaw, who was one of the few attendants he could trust. A large middle-aged woman with a soft voice and a calm, open face, Shaw would at least listen to him before calling the administration or the security office.

As he opened the door to the corridor, Fred saw Miss Shaw talking with two boys. She glanced his way but registered no surprise at seeing him. He sidled up to her, and she put her hand up and said, "Now wait just a minute, Freddie. I'm talking to these boys now."

When she was finished with the others, Regina Shaw turned to Fred and noted that he had been giving a few people at Fernald "a run for their money." Fred smiled and confessed that this was the case. But he insisted he had good reason to run away in the first place, and to fight his placement in the North Building. He was not retarded. Putting him in North signaled to everyone in the institution that he was. It was insulting to him, and he wouldn't stand for it. "They can't put me back there," said Fred.

"I don't think they will, since you already proved you know how to get out," she responded.

The problem of a runaway, who cannot be contained, was too big for any attendant to handle on her own. Shaw told Fred that Dr. Lawrence Kelley would be at the BH within the hour. Fred could wait and tell his story to him.

Dr. Kelley had examined Fred and signed his admission papers—confirming he was "feebleminded of the familial type"—nearly ten years before. As far as Fred was concerned, he was a cold and distant man, oblivious to the true character of the boys he treated. Fred believed the doctor consulted each patient's file to see whether he was a moron, imbecile, or idiot, and then dealt with him according to that diagnosis. For this reason, it would be up to Fred to force him to see something more.

Regina Shaw brought Dr. Lawrence Kelley into the room where Fred was waiting, and lingered to answer a few of the doctor's questions about Fred's disappearance, capture, and subsequent escape from the North Building. Kelley seemed to be impressed by what he heard about the boy's resourcefulness. When Miss Shaw got to the part about Fred getting out of the cell, he interrupted to ask for more detail. He was amazed to hear that Fred had made a hammer and chisel, and had also enticed Mr. Bell to open the door.

When Dr. Kelley finally turned his attention to the boy in question, Fred had his plan firmly in mind. He answered a few basic questions, about how he had gotten to Dracut and what he had done there. When the subject turned to his behavior at the North Building, Fred said he had felt furious because he had been placed in an infamous dormitory, a wretched place that everyone at Fernald identified with "retards."

When the doctor paused in his questioning, Fred then seized control of the conversation. Decades later, he would recall what transpired.

"Can I ask you, Dr. Kelley, are you Catholic?"

"No," came the answer, and for a moment Fred felt derailed. Then Kelley continued, "But I do believe in God, the same God as you."

"Then I want to ask you this question. If you promise, if you swear on God that you'll tell the truth, then I'll go back to the North Building if you want me to. I won't fight it. But you've got to swear to tell me the truth."

"All right," said Kelley. "I swear."

Fred paused for a moment to get the words just right. Then he stared at the doctor and spoke very slowly. "If you had a kid, and even if that kid was retarded, for real, would you put him in this place?"

Kelley didn't answer right away. When he finally started to speak, he sputtered a bit and then stopped. Finally, he turned to Regina Shaw and asked a question of his own. "What's this kid doing here?" She shrugged. Then the doctor turned back to Fred.

"I wouldn't put my child here," he finally confessed. "You're right. I wouldn't do it."

Lawrence Kelley's impression of Fred Boyce is evident in the report he wrote after their meeting. In it, he recounts the escape, Fred's capture, and the events at the North Building. He notes that he spent two hours with Fred, and then makes a strong argument for his immediate training and release to the outside.

"He has much merit in his philosophy," writes Kelley. "I believe he is ready for a new start and will make good if given help and encouragement." He argues that Fred is a solid candidate for the social workers' new training program, and notes that Fred has the IQ "to back it up." With a tone so poignant that it seems clear that Fred got through to him, Kelley pleads with those who might read his comments.

"Patient is now at the crossroads," he concludes. "We should not fail him at this time."

With Kelley's support, Fred was allowed to stay in the BH. He met with a psychologist who found that "though he is somewhat immature intellectually, he is certainly not feebleminded." Given that he had planned and executed his escape from Fernald and then found both a home and a job, this was a fairly obvious fact.

In one of the tests the psychologist gave Fred to gauge his abilities— a quiz on the meaning of various proverbs—his score fell in the range of the average adult. "The examiner had tested well over 500 patients in this school," wrote the psychologist, "and this is the first time that he can recall anyone succeeding on this particular sub-test."[2]

The observations of a psychiatrist who found Fred ready for a "new start" and a psychologist who declared him "not feebleminded" should have made him a solid choice for training and parole. But in the aftermath of the riot, and the long period of unrest that preceded it, many of the so-called borderline cases at Fernald were being reconsidered. So many of the State Boys were discovered to be not-so-feebleminded that their numbers overwhelmed the still-new program for training and parole.

The Gagne brothers were examples of this phenomenon. By the start of 1958, Robert's IQ was suddenly 80 points, the highest it had ever been. Albert, who had been selected for the release program, was doing so well in his training and counseling sessions that it was time for his transfer to Emerson Hospital, on the outside.[3]

It happened during the time when Fred had been away in Dracut. Albert had reported for work at Dr. Benda's laboratory to find that the doctors and technicians had brought a cake into the office. They had gathered to wish him well as he left Fernald for a job and a home on the outside. It was a complete surprise to Albert. He knew he was being groomed for such a move but had not been told that the day was very near.

As soon as the cake was consumed and a few speeches had been made about what a fine boy he was, Albert walked to the Administration Building for a meeting with Malcolm Farrell. The superintendent, imposing in his suit and tie, brought Albert into a conference room where they sat together at a big polished table. Albert trembled a bit and sweated a great deal as Farrell said that he expected him to behave like a gentleman and make the school proud. When Farrell finished and showed him the door, the one word Albert remembered was "rehabilitate." Dr. Farrell had said that parole would "rehabilitate" him.

Abigail Bacon was waiting in the hallway when Albert emerged from the superintendent's conference room. She was holding a pair of trousers, a shirt, shoes, and a jacket. These were, in State Boy vernacular, "outside" or "downtown" clothes. He was to put them on right away.

After he was changed, Ms. Bacon drove Albert to Concord, where she had arranged for him to wash dishes at Emerson Hospital and live in a dormitory on the hospital grounds. He would be paid less than minimum wage, and the money would go into a savings account that he would not be allowed to control. His comings and goings would be monitored. But if he did his work, kept his room clean, and got along with others, he could ultimately earn full independence.

Unlike Joey Almeida, Fred Boyce, and others who bristled at the term "parole" because it suggested they were criminals, Albert Gagne embraced it. He didn't care what it was called, or that he was going into another sort of institution where freedoms would come slowly. The only thing that bothered Albert was that he would have to leave his younger brother Robert behind. This troubled Robert, too. In the months after Albert departed, staff would describe Robert as "grave-faced" and "indifferent."[4]

If Robert was jealous and resentful of the opportunity given to his brother, he wasn't alone. Most of the young men in the BH believed they didn't belong in an institution for the retarded, and a great many of them were correct. Angry, and filled with the energy of adolescence, they were impatient with teachers and attendants who told them their turn would come, eventually.

In a report to Fernald trustees, principal Mildred Brazier noted that children often arrived at the school from their dorms emotionally upset, "screaming, quarreling and abusive to each other." She asked if "overcrowded, poorly guarded dormitories, enforced inactivity and lack of motivation" contributed to the "children's difficulties."[5]

By the spring of 1958, only seven of the fifteen Ward 22 rioters had returned from Bridgewater. The state's determination to keep the rest locked away was a reminder that, despite the kindness of new young social workers and teachers, protest could have a terrible outcome.

As one of the lucky seven who had been bused back to Fernald, Joey Almeida was still paying for his role in the uprising. He had been confined to a new locked ward—referred to as the temporary Ward 22—which was located in the Wallace Building. Joey was permitted to return to manual training and to school. Now sixteen years old, and a seven-year veteran of the Fernald system, he was more certain than ever that he belonged on the outside.

On March 24, 1958, Joey spent a couple of hours with the young teacher Lawrence Gomes. Well liked by the State Boys, Gomes was part of the new generation of teachers at Fernald, many of whom believed that their pupils had more potential than was previously recognized. In Joey, Gomes encountered a teenager who, other than for the effects of being a resident of Fernald—anxiety, nervousness, low education—seemed quite normal.

Like many normal teenagers would, Joey first resisted Gomes's attempts to get him talking. He answered in grunts and small sentences filled with slang. He described certain attendants and boys in the BH ward as "crumbs" and "queers." Then, as he began to trust the teacher, the truth behind these words became clear. Joey was again being pressured into sexual acts by both the staff and bigger, tougher boys.

As soon as he confided in Lawrence Gomes, Joey begged him to keep what he had revealed a secret. He insisted that nothing good would happen if Gomes reported it to his higher-ups. And he said that his tormentors could make his life even worse if they thought he had squealed. Gomes said he wouldn't act on what Joey had said, and this sealed the bond between them.

Afterward, Gomes wrote a report damning the institution for what had been done to Joey and predicting a troubled future for him:

> I myself find no evidence that he is significantly retarded, particularly to the degree that requires keeping him in the school. All Joseph needs is someone who is able to counsel him outside the institution, to have close contact within a normal setting, which we do not have.
>
> We shall never know what he could have been had he not been a victim of a sincere but misguided system. The tragedy here will be one of missed opportunity. Sadly to say, this can never be regained. Once Joseph has gone from this institution the stigma will follow him for a long time. His injuries you won't see, because they will be injuries of the mind. There will be a time when he will be preoccupied with destroying his past, when most are preoccupied with building their future. Some kind of effort should be made in helping with a program to get Joseph out of Fernald. It should be discussed at considerable length with the superintendent and doctors of this institution.

Gomes understood that Joey was not unique in his experience at the school. Sexual encounters between staff and residents were common, and the boys all knew which attendants were predatory and how they tried to get what they wanted. Some used threats and physical pressure. Others bribed boys with candy. Sexual activity among the boys was also widely acknowledged. However, not everyone was as open as Joey when it came to discussing it. Fear, shame, and worry about the stigma of homosexuality kept many quiet. Some boys may have also been so traumatized that they had trouble admitting what had happened to them. They could agree that yes, sexual abuse was a problem on the ward, but they would never complain on their own behalf. Finally, there were boys who discussed the problem among themselves but would never bring it up to a teacher, counselor, or administrator. They were sure nothing would be done.[6]

Although it's not noted in what he wrote, Lawrence Gomes was undoubtedly inspired to outrage by something more than his encounter with Joey. On the very week that the two got together, *Life* magazine published a lengthy article that surely caught the attention of every professional in the field of mental retardation. Its focus was a tragic figure named Mayo Buckner.

In 1898, eight-year-old Mayo had been taken to the Iowa Home for Feeble-minded Children by his mother, who was alarmed because he was extremely shy, ate fast, and sometimes rolled his eyes like Blind Boone, a vagabond entertainer she had seen while pregnant with him. The school superintendent, who said that he assessed children by intuition, accepted Mayo as a "medium grade imbecile" and listed the cause as the "prenatal influence" of Blind Boone.

(The idea that a mother's emotional experience could affect a fetus in her womb, and somehow cause feeblemindedness, was not unique to Mayo Buckner's case. In the early twentieth century, many experts embraced this notion, and it persisted for decades. Well into the 1950s, the official Description Application for Admission to Fernald and other institutions in Massachusetts included the following question: "During the mother's pregnancy, was there any illness or injury, or any unusual emotional disturbance, anxiety, or fright?")

In his early adult years, Mayo Buckner had tried to persuade school staff that he was normal. He had begged his mother to be freed. These pleas always failed, and gradually he became accustomed to life in what writer Robert Wallace described as a place that leaves a visitor feeling "that it is always cloudy and about to rain." Mayo became a fixture within the institution and even enjoyed a romance with an inmate named Valencia, who was also of normal intelligence. Their relationship lasted ten years, and in those rare moments when they were alone together, he held her hand.

Sixty years after Mayo was admitted to the Iowa school, a young, newly installed superintendent named Alfred Sasser had him tested and discovered his IQ was 120. He offered him immediate release, but

Mayo declined. He had become a highly valued worker at the school, which gave him a feeling of community and competence. He doubted whether he could adjust to life outside.

Mayo Buckner's humble acceptance of a life that had included whippings, deprivation, and isolation made the tragedy of his confinement vivid. The *Life* writer's estimate that at least 5,000 Mayo Buckners resided in institutions nationwide—a figure that was later found to be conservative—made clear the size of the scandal that he represented.

Some *Life* readers responded to the Mayo Buckner article with outrage, but it was short-lived. No one organized a campaign on behalf of the tragically misdiagnosed children and adults who lived in virtually every state school in America. In contrast, by the time the *Life* piece was published, the National Association for Retarded Citizens had roughly 30,000 members and was making great gains within institutions and with government officials. Local parents groups within the NARC had developed more than 1,000 programs for their children, and the national organization had made retarded children a popular cause.[7]

Because they were orphaned or rejected by their families, institutionalized "borderline cases" like Fred, Joey, Charlie Hatch, and others had no one to speak for them and were invisible to the outside world. However, through the riot at Ward 22 and their many personal rebellions, the State Boys had made themselves a very real problem to those who ran Fernald. Some of the school's new social workers and teachers, who were actually paid by programs that had been created in response to the NARC advocates, recognized their plight and intended to fix it. The pace of change would be agonizingly slow for the teenagers who found every day at Fernald an insult, but it was coming.[8]

EIGHT

Like Albert Gagne, many of the State Boys were getting smarter. At least that's what their IQ tests showed. In 1958, one boy after another hit a record high on his annual exam. Fred Boyce, who was once found to have an IQ of 65, suddenly scored 91. Joey Almeida's IQ measured 93. If these increases had had something to do with their education, the staff of Fernald would have been rightly proud. But the boys were receiving almost no education, and what little they did get was simply a repetition of the grammar school lessons they had received many times over.

The real reasons for the higher scores could have been found not in the boys but in the test and the people giving it. By the late 1950s, Fernald had finally adopted the new Wechsler Intelligence Test, which measured intelligence as it relates to a person's ability to reason and solve problems. At the same time, a new generation of professionals had arrived at Fernald. These teachers and psychologists were better at getting the most out of each person they examined by making them feel comfortable and relaxed.

"We knew that the tester had a lot to do with the scores," recalled Lawrence Gomes, decades later. "We also knew that IQ didn't tell you everything. I mean, you could just talk with some of the students, and you knew they weren't really retarded and that there was some other reason why they were in the place."[1]

Gomes was the teacher who had believed Joey Almeida's reports of sexual assaults and recommended, in vain, that he be helped to leave Fernald as soon as possible. Unlike others on the school staff, who thought their students were limited by their lack of "gray matter," Gomes suspected that for most Fernald students their diagnoses— *moron, idiot, imbecile*—had become a self-fulfilling prophecy of failure. This was especially true for the brighter Fernald students, who he believed were victims of the state's indifference. "There was no body of knowledge, nothing you could point at to prove it," he said. "But if you just looked at these kids, you knew that it was true."

Actually, the body of knowledge that would have proven Gomes's point, and changed the course of countless lives, had been available for decades in the files of Fernald and in the records of other state agencies. But no one had even proposed to look for it until 1958, when the National Institute of Mental Health gave a young psychiatrist named Edward V. Jones a grant to begin compiling it all.

Jones was interested in understanding why certain boys and girls who were deemed "borderline" retarded wound up in institutions, while others did not. In his grant application, he wrote that he was especially curious about the "deprivation of essential physical and emotional needs of prenatal and early childhood." He added that "the question of heredity, while possibly a contributory factor, is not easily studied and will not be part of this research, or at least will be held in abeyance for the time being."

Once their work got under way, Jones and his collaborators identified 250 of Fernald's higher-functioning children and then searched their records for what they called "predisposing factors" to their admission. They found that all but one of these boys and girls came from a "disorganized" family affected by problems including alcoholism, poverty, and unemployment. (One-third of their fathers, and 41 percent of their mothers, had been charged and found guilty of child neglect.) All but three of the 250 children had been involved with social service agencies, and most had behavior problems reported by parents or teachers.

Digging deeper, the study group eventually found that many of their subjects had never attended school at all, and that Fernald's records grossly underestimated the number of foster homes they had occupied. (The disruption caused by frequent changes of homes and guardians was known at the time to cause serious psychological and educational problems for children.)

Jones's findings were provocative, especially given the times. For example, he discovered that African-American children were underrepresented at Fernald because "not much is expected of a Negro child" in the first place and therefore most escaped official attention. He also discovered that the longer a child managed to stay in the community, the less likely he was to be institutionalized at all.

But the most dramatic of Jones's discoveries was data which showed that thousands of Fernald children had scored low on IQ tests because of the state's negligence. Fred Boyce was a perfect example. He had become a ward of the state in infancy. From that moment on, every decision made on his behalf pushed him further from the basic education he needed and closer to institutionalization. Considered in this way, the findings of the Jones group added new meaning to the term State Boy, placing the responsibility for his fate firmly on the state, and not the boy.[2]

Despite all that conspired against normalcy in their lives, the young men of the Boys Home continued to assert themselves as regular, 1950s-style American teenagers. When they were allowed to control the dial, the radios in the dorm were tuned to early rock 'n' roll outlets. The boys' favorite disc jockey was Arnie Ginsberg of WBOS, who used bells and horns and whistles in a nightly Top 40 program. Ginsberg's chief competitor, George Fennell of WHIL, matched him stunt-for-stunt. Once he locked himself in his studio and played Little Richard's "Tutti Frutti" over and over again for nearly a full day.

Boston clergy had called for a ban on rock programs, and a psychiatrist in Connecticut had described the music as a contagious disease. But for the State Boys, the songs and the jockeys expressed rebellious feelings and thoughts the boys were forbidden to express themselves. Every chance they got, they lay on their beds and soaked in the melodies and lyrics.[3]

The urge to share the interest of other teenagers in America reflected an emotional shift for many of the boys. Like most adolescents, they set aside some of their childhood desires to stand out, to be truly known and accepted as individuals, and embraced the styles that marked them as part of a group. Many of the State Boys wanted nothing more than to merge with the images they saw on *American Bandstand*, to be indistinguishable from normal boys their age.

Just as basic to their identities as young men was the way many of the boys in the BH obsessed about young women. Some of their talk may have been the product of the anxiety and shame associated with the homosexual activity that was so common in the institution. But whatever the reason, they talked about girls constantly, and flirted with them whenever they got the chance. At the handful of dances held each year, they competed fiercely for female attention. The same thing happened at church services and movies.

The dances were peak moments of excitement, and a rare chance to feel normal. Boys would be allowed to wear outside clothes, including secondhand jackets and ties that were reserved for special occasions. Girls put on the dresses they normally wore to church. It didn't matter that their outfits were mismatched or outdated. For an evening, they felt like regular teens, posing on the riser in the gym, trying out dance steps, and fantasizing about kisses and more.[4]

Still imagining that Margaret Burney was his girl, Fred tried to impress her by always dressing as well as he could. Every day, he searched through the piles of shirts and pants in the BH clothes room to find the ones that fit snugly and seemed newer. In the spring of 1958, he was thrilled to learn that he had received $102 from the Veterans Administration—a long-delayed survivor's benefit related to his father's death—and would be allowed to spend it any way he chose. He chose to go shopping for clothes.

Of all people, it was James McGinn—the spoon-wielding terror of the BD—who arrived on a Saturday to take Fred on his shopping excursion. McGinn had a car, and he had volunteered to drive Fred to downtown Boston. He took him to department stores and small shops where he pushed him to buy conservative shirts and trousers. Thrilled, and a little confused by the vast selection, Fred allowed McGinn to steer him.

At the end of the day, McGinn turned serious and told Fred that he was disappointed in him, because he hadn't offered to buy dinner. Startled into the habit of obedience that he developed under McGinn at the BD, Fred quickly volunteered to spend the cash that remained for a meal at a restaurant of McGinn's choice. That night, McGinn returned to Fernald well fed on a steak dinner. Fred went back to the BH with his money all spent.[5]

Nice clothes looked best on sculpted bodies, and some of the residents of the BH adopted Jack La Lanne, Joe Weider, and George "Superman" Reeves as bodybuilding role models. Eric Johnson and his friend Jerry Clements became even more dedicated weight lifters. They turned a basement room into a small gym and devoted countless hours to lifting and pushing weights around. Once their bodies became big and strong, the institution put their muscle to work, assigning them to jobs involving lots of heavy lifting. Johnson spent much of the winter months shoveling coal into the huge furnaces that created steam heat for the school. His chest and his back would be permanently scarred by the burns he received from the red-hot cinders that occasionally flew out of the fire.

The burns were worth it to Johnson, because his hard labor brought respect from both his peers in the Boys Home and the staff at Fernald. He was frequently recruited to perform heavy work. One assignment involved lifting iron fire hydrants and putting them in place along a newly installed water line. He didn't resent this work at all. Instead, it made him feel competent, valued, and even superior to the Fernald staff.

The jobs were vital to the boys' self-esteem, especially since they had been denied the chance to achieve in school, or sports, or any of the other normal ways a teen might distinguish himself. The State Boys were not told about their rising test scores. Nor were they informed of the research on their records. Most of the attendants continued to call them morons and treated them as if they were doomed to be lifelong wards of the state. Nevertheless, they understood that a few, such as Albert Gagne, had gotten out. And the new social workers and psychologists talked openly about jobs on the outside and places where they might live.[6]

The politics of the day had also begun to move in the State Boys' favor. On a national level, the Kennedy family had created a large charity that had begun to make grants for the study and treatment of mental retardation. The giving was motivated in part by political ambition—it allied John F. Kennedy with an appealing cause. But the matter of Kennedy's sister Rosemary, who was lobotomized, also lay behind it.

The Kennedy funds attracted the interest of Dr. Benda, Fernald's research director, who met with the charity's overseers, Eunice and Sargent Shriver, in the fall of 1958. In a follow-up letter, Benda argued that the Fernald School would be a premier site for a large research facility. As a second option, he suggested a new three-hundred-bed hospital for the temporary care and assessment of babies and young children whose parents did not want to relinquish custody. More than twelve years would pass, and Dr. Benda would retire, but eventually a center like the one he described would be opened at Fernald.[7]

While the Kennedy family and others such as the National Association for Retarded Citizens worked on a national level, politicians in many states also pressed the cause of the retarded. In July 1958, a published audit in Massachusetts documented the gross overcrowding and dangerous understaffing at state institutions. Given advances in medicine, the report warned, these problems were going to become worse as more severely disabled children would survive to need more intense care. Administrators at Fernald who contemplated such demands surely saw that they needed to make room.[8]

At the start of 1959, as promised by Dr. Kelley, Fred Boyce was put on the program to prepare him for release. His training would consist of a new job moving boxes and bulk items around the Fernald warehouse and weekly group therapy sessions with a social worker named Mr. Avey. It was the first time that Fred or the others in the group had received any regular attention from a mental health professional. As might be expected with half a dozen teenage boys who were sensitive to the stigma of psychology, progress was slow.

Fred thought that he didn't need counseling, and the proof was the success he had in Dracut, where he had spent almost ten days on his own. When the therapist discounted this experience and said that Fred couldn't possibly know what faced him on the outside, Fred refused to back down. He would depend on the new friends he would make for help, he said. He said the counseling was "foolishness" that only made him feel worse about himself.

Many of the State Boys resented contact with the therapists—they called them "psychs"—because they felt it signaled their acceptance of the label "retarded." They didn't trust psychs, and this anxiety was made worse during a session when Mr. Avey left the room and Fred found a small tape recorder hidden in an arrangement of plastic flowers. When he realized what it was, he held a finger to his lips and quieted the others. He then leaned over, so he was close to the machine.

"Quack," he said, letting the sound flatten in the back of his throat. "Quack, quack, quack!"

He stopped, looked at the others, and they all started grinning, and laughing, and quacking, too.

"Quack!"

"Quack!"

"Quack, quack!"

The noise brought Mr. Avey back into the room, where Fred and the others confronted him with the recording device. He struggled to explain why he was taping them without their permission but would never recover the trust that had been lost.

After the session that ended in all the quacking, Fred began meeting privately with a new psych, a social worker named Alessandrini. Alone with Mr. Alessandrini, Fred eventually opened up. Records from those sessions reveal a very anxious, angry young man.

". . . a great deal of criticism and hostility toward the institution and its personnel," noted Alessandrini after one session. "He has a great deal of feeling about his being placed at Fernald, and it is difficult for him to associate any pleasant thoughts with his eleven years of institutional living."

With Alessandrini, Fred became more open about his feelings. He spoke in mournful terms about his time at Fernald as a "terrible waste." He described a string of incidents in which attendants beat and humiliated younger children. (He preferred to talk about the indignities suffered by others, not his own.) In his notes, the social worker described Fred as "hostile" and "resentful." In one meeting, Fred let down his guard enough to cry as he explained that since coming to Fernald he had never experienced "feelings of belonging, or that anyone cared."[9]

*   *   *

Joey Almeida followed Fred into the preparole program. He went back to work at the Fernald bakery, which produced everything from the doughnuts laid out for staff meetings in the Administration Building to the loaves of plain white bread served in the school's lunchrooms.

Recently renovated, the bakery was one of the most modern facilities on the campus. The bread-making process was highly mechanized. Dough was prepared in a huge mixer, and once it was mixed, automated machines cut it into loaves. Once the bread was baked, more machines did the slicing and packaging. Joey then slid large trays of the springy white loaves onto racks that were then wheeled onto trucks for distribution.

The bakery was bright, clean, and airy. Joey liked the clean white aprons and hats he wore for every shift. He loved the smell of flour and yeast and the warmth of the ovens on cold winter days. The men in the bakery treated him well, especially master baker John Sullivan. They let him sample the sweet products they made and praised his efforts. He began to think that of all the jobs a boy could do at Fernald, his was the best preparation for the outside. The world would always need bakers.

Notes written by Joey Almeida's counselor show that, like Fred, he was also certain that he didn't belong at Fernald. However, he was more determined than Fred to curry favor from those who ran the release program. He was terrified by the prospect of somehow being sent back to Bridgewater, where he knew some State Boys remained locked away, and he took every opportunity to demonstrate that he was a cooperative and capable boy. In one therapy session, Joey insisted he was an advanced reader and demonstrated it by reading the titles on the spines of the books behind the social worker's desk. At another meeting, he said he had learned to get along at the Boys Home without fighting every time someone said something he didn't like. He even allowed that he had gotten some benefit out of Fernald.

But as much as Joey intended to please, he sometimes forgot his goal and taunted his psych. He said that everyone in the program told lies in order to please the adults who controlled their freedom. And he frequently returned to his old attitude, insisting that he didn't belong in the school and that the place had done him enormous harm. He was fed up with Fernald and desperate to get out.

Joey was not the only State Boy who struggled with his desire to please, so he could be released, and his anger over his years of confinement at Fernald. Though they dutifully attended meetings with psychs and were diligent in their work, many boys also rebelled when they got the chance. They continued to break into buildings after hours, robbing vending machines and stealing food. If he was especially bored, or angry, Joey would join one of these raiding parties and get some satisfaction from defying the institution.[10]

The Boys Home staff was not up to the job of policing every resident every hour of the day. The State Boys were just too resourceful and too numerous to be monitored that way. Recognizing this, a new doctor at the dorm came up with a scheme to put the onus on the boys themselves. Adrian Blake, M.D., proposed a self-governance scheme in which the group elected leaders and imposed penalties on rule-breakers. In exchange, the boys were to be granted certain privileges, the most important being permission to walk the mile or so to Waverley Square and its stores, coffee shop, and soda fountain. Blake took responsibility for this project and agreed to answer to Malcolm Farrell if anything went wrong. But on a day-to-day basis, Regina Shaw, the same softhearted nurse who had helped Fred after his flight from the North Building, oversaw it.

Blake, who had emigrated recently from Ireland, had been disturbed by the presence of so many able boys at the BH. "Most were normal kids who were bored," he explained decades later. "But some were really headed for trouble. Earl Badgett was one example. He was the one found with a gun in his possession. I was always worried that when he got out that either he would kill someone or someone would kill him."

By giving the boys of the BH more freedom, but tying it to self-control, Blake hoped to establish a peace that had eluded the staff for years. As further inducement, he told the State Boys that they could fix up an abandoned outbuilding near the BD as a clubhouse. They could use it for meetings and socializing.

The Blake project, which everyone came to call Boys Town, appealed to the residents and they immediately organized elections.

Fred Boyce became president in a landslide vote. His first serious duty involved disciplining his best friend, Joey, who had run away, only to be brought back to Fernald by the Boston police department. Visits to Waverley Square had been suspended pending the outcome of the case and the return of order to the BH.

Acting as a judge, Fred conducted a mock trial inside the club-house. Half a dozen boys from the BH sat as a jury while Joey argued that he had been coaxed into leaving by other boys. He promised that he would never do it again and then begged to be found not guilty. Joey was very persuasive. And despite all the evidence against him, the jury found him not guilty. Joey threw up his hands and the crowd of boys cheered, but the victory was short-lived. Fred hammered the table with a gavel, which Nurse Shaw had given him as a token of her confidence in his leadership.

"The judge is overruling the decision," he said. "Joey did it, and he is sentenced to mop the annex at the BH every day for a month."

With Fred's ruling, and Joey's willing acceptance of his sentence, the vandalism, escapes, and fights stopped. Waverley Square privileges were restored. As further reward, the school's head gym teacher, Joe Barnes, organized a baseball game with a local team. It would be a chance for the State Boys to test themselves against the outside.[11]

Over the years, many of the details of the game would become blurred. Some would recall that a team of boys from Belmont had come to play. Others thought the opponents were from Waltham. But those who were there would agree that the outsiders were wary at first and stared at the Fernald boys as if they were expecting to catch them behaving strangely. Once the game started, they stopped staring, because the Fernald boys played as well as they did.

In the course of the game, Joey Almeida's brother Richard made a series of saving plays at shortstop. Joey and Fred played well in the outfield. Pitching from a full wind-up, Robert Gagne struck out many of the visiting hitters. But the strongest efforts were made by the bigger, older members of the team—Eric Johnson, Jerry Clements, and forty-year-old Charlie Dolphus. Their hitting won the game for the Fernald team, but the score wasn't the important thing for the State Boys. Their satisfaction came with seeing the surprise in their opponents' eyes.

That was better than spending time off the grounds in Waverley Square.[12]

A few weeks after the game, Robert Gagne was paroled to join his brother Albert at Emerson Hospital, where the two were allowed to share a room. In his time on the job, Albert had done so well that the kitchen supervisor had assigned him to train new employees. He had received a commendation for extra work done during a storm that caused flooding in the hospital. And he had made a close friendship with a newly arrived Hungarian refugee who also worked in the kitchen. Albert had helped the newcomer learn a bit of English as well as the routine of the kitchen.

The Gagnes' adjustment to the outside was not entirely smooth. One major problem began with a bike trip to a shopping center near the hospital. There, Albert and Robert discovered a shop that sold televisions, radios, electric guitars, phonographs, and amplifiers. When they learned they could buy these items with a few dollars down, and extended monthly payments, they bought one of each, and threw in an accordion for good measure.

Electronic and musical gear filled every corner of Albert and Robert's room. When she came for a visit, Abigail Bacon questioned them about how they had acquired it all. After they told her, she spent more than an hour explaining the trap they had fallen into. With her help, the stuff went back to the store. The money they had paid was refunded, and the lesson taught by this experience was reviewed more than once.

Once they felt less embarrassed by the financial debacle, Abigail Bacon began to press the young men she had placed at the hospital about keeping their quarters clean. It took a few months for her to get them to organize the chores and establish a routine that got them done. But once they got the hang of it, they almost seemed to enjoy it. "The floors were waxed, the bathroom was clean, the beds were made with clean linen," reported Ms. Bacon. "They all said it was more comfortable to live in orderly quarters."

By the summer of 1959, Ms. Bacon was referring to the Gagnes as the "Emerson Hospital boys" and describing the "increasing maturity on their part." Albert and Robert went to Hampton Beach, New

Hampshire, for a summer vacation, staying two weeks in a rented cottage. In October 1958, Albert was formally discharged from the state's care and declared "capable of self support." He had saved a small fortune—$491.20—and the only worry expressed by Ms. Bacon was the fear that others might take advantage of him.[13]

For the State Boys left behind, Abigail Bacon's reports on the success of the Gagne brothers was a source of hope. This hope was matched, however, by the fear that rose in their minds every time they thought about the Ward 22 rebels who had yet to return from Bridgewater. The two-year anniversary of the uprising was fast approaching, and yet the fates of those eight boys remained a mystery. Whenever Fred asked an attendant, or nurse, or doctor about the missing boys, his questions were shrugged off. He was told to forget about them and make sure that he behaved well enough to avoid Bridgewater himself.

In the interest of reaching their goal—release from Fernald—Fred and some of the others may have been able to push the missing eight out of their minds. What good did it do, they asked themselves, to dwell on something they couldn't control? Bobby Williams, however, could not stop worrying about his twin, Richard, who was among them. The separation from his brother was a source of constant grief for Robert. He felt like he had lost a limb.

Richard, who now had a second state number—1414—identifying him as a prison inmate, sent several letters. Each was a mixture of advice, promises, and pleading.

Dear Brother Bobby,

I want to tell you I am doing everything I am told so I can get my release from this dept. and I want you to do everything you are told to do up there.

It is for your own good because if you do what you are told you will get [a] better chance of getting out.

. . . will you please write me a letter, for I would like to hear from you Bobby, and when you get out from the State School will you come down here to see me Bobby[?].

February 18, 1958

Richard's letters were eventually followed by formal communications from Charles Gaughan, superintendent of the Bridgewater prison. On May 20, 1959, a year and six months into Richard's incarceration, Gaughan wrote to Malcolm Farrell suggesting that the twins be allowed a visit. Another state school for the retarded had allowed such contacts, he noted. "Naturally, if such a visit is possible, much will have been done to improve the morale of both of them."

Gaughan's request was taken up at a Fernald staff meeting and unanimously rejected. On May 26, Dr. Farrell wrote, "It was felt because of Richard's condition and chronic troublemaking propensities that it would be better that his brother Bobby not visit him. Bobby has been doing quite well here since his brother has been at your institution and we all feel that he would be better in every way the less contact he has with his brother."

Bobby was not told about the prison superintendent's letter. Nor was he consulted about Malcolm Farrell's reply. But in his own way, he made it clear that he did not agree that it would be better in every way for him to remain apart from his brother. On July 15, he ran away from Fernald, intending to get to Bridgewater to see his twin. He left behind a note:

Dear Sir,
    I will not be back any more. If you want me you better try and hunt for me.

From,

BOBBY WILLIAMS
    The cops will never catch up to me. Ha Ha Ha Ha

The cops caught up to Bobby Williams the very day he ran away. He was allowed to go back to the Boys Home but was warned he would be placed in an isolation cell if he continued to rebel. A week later, while working in the kitchen of a ward for blind children, Bobby began waving a dull knife at one of the cooks. Attendants would say he had a wild look in his eyes and seemed to have lost control. As punishment, he was sent to the renovated Ward 22, and then transferred out of the Boys Home and into Dowling Hall, a building where he would be isolated among older, truly disabled men.

Years would pass with Richard remaining in Bridgewater. In that time, Bobby, who was separated from the other bright boys, heard vague reports about how, one by one, they were entering the parole program. His life was reduced to the ward and day room at Dowling Hall and the kitchen where he worked. Eventually, all of Bobby's friends would depart, and he would feel ever more forgotten and abandoned. He continued to fret over Richard, and to dream about helping him.[14]

In reality, if Bobby had had the freedom and the wherewithal to act on Richard's behalf, he would have met enormous obstacles. The relatives of a few of the rebellious State Boys were trying to get them out of Bridgewater but were having no success. Charles Hatch's older brother Edward had read a newspaper article about the Ward 22 riot and immediately contacted a lawyer to help his brother. He paid to have three psychiatrists examine Charles, and each one certified that he was not dangerous and argued for his release. State officials were unmoved, and Charles Hatch would remain at Bridgewater serving as an example of what could happen to the State Boys of Fernald if they strayed too far.

Hatch was remembered at the Boys Home and by those who were paroled to live on the outside. Rumors about him—that he was dead or was being tortured—ran rampant. In reality, Charlie was safe at Bridgewater, but it was a miserable existence. Older inmates often forced him into fights, and his so-called rehabilitation required that he take medications that made him feel half-asleep. He received no education, not even basic lessons on the demands of everyday living, such as managing money.[15]

After Fred Boyce had spent a full year moving boxes and bulk items around the Fernald storeroom, his supervisor was satisfied that he knew how to work with others and handle the responsibilities of a job. (Joseph Ready, who was in charge of the storeroom, made special note of Fred's facility with "pencil, paper, and marking crayon.") At the same time, Boys Home attendant Regina Shaw reported that Fred got along very well with both his dorm mates and the staff. She added,

Fred "had been well able to take care of himself right along." His psych said Fred had been "almost friendly" in recent months and that he was a "promising boy who should do well in community placement."

On February 9, 1960, a panel of doctors, social workers, and psychologists agreed that Fred, who had reached age nineteen, could be trusted on the outside. He would be paroled under supervision, which meant they could bring him back for almost any reason and without a court hearing.

Unlike the Gagnes and others who took jobs at institutions, Fred wanted nothing to do with big brick buildings, government workers, or medical environments. He told his social worker that he wanted an ordinary job, working with ordinary people, and a room of his own in a regular house.

Ten days after his parole was approved, Fred went for a job interview at the Griffith Ladder Company, a small manufacturing company that was less than two miles from the Fernald campus. Four days after that, he reported for work. At first, he was assigned to use a hammer and nails to fix the rungs in wooden ladders. Soon he was able to place the nail and hammer it flush with a single, well-leveraged blow. He worked faster than the assembly line. When the plant manager noticed, Fred was transferred to run a riveting machine on the line that made metal ladders. He was paid $1 per hour.

During the first month of his parole, Fred continued to live at the Boys Home, walking to work each morning and returning at night. Though he was excited about being freed, he had to prove that he could manage work before he could leave the dorm. Finally, on March 11, he was told that a space had been found for him in a private home on Harris Street, in downtown Waltham. (The rent was $1 per day, which would come out of Fred's earnings at Griffith Ladder.) A social worker came to the BH and gave him some paper bags so he could gather up the clothes he had bought on his shopping trip with James McGinn.

Since he had known for weeks that he was going, Fred had already said goodbye to Joey Almeida, Charlie Dolphus, Eric Johnson, and the others who had not yet been paroled. When he packed his things, they were all out on the campus anyway, so there was no one to watch him leave, no one to bid him good luck. He took one last look at the rows of beds in the ward, turned his back, and went downstairs to where a car waited to take him to Harris Street.

Though Fred's release from Fernald lacked outward drama, it was a highly emotional moment. He was both apprehensive about being on his own and desperate to succeed. Most of all, he dreaded the humiliation of having to return to the BH if he somehow failed to cope on the outside.

At the Harris Street house, the landlord showed him the private outside entrance to his room. Fred put away his clothes in the dresser and said goodbye to the social worker. When he was finally alone, he sat on the small bed and experienced, for the first time in his life, a sense of privacy and independence. He wasn't entirely comfortable.

In the months that followed, Fred fell into a routine that was nearly as rigid as the one that had ruled his life at Fernald. He awoke early enough to catch a bus to Belmont for an eight o'clock start at the factory. (This was a bit of a problem on his first day because he had never actually ridden a city bus. He waited at the stop, but when the bus arrived, he just stood there waiting for the driver to invite him to board. Everyone else got on, and the bus left. A man who had seen Fred miss the bus told him to stop being so polite. He got on the next one, and was never again late for work.)

Once he arrived at the ladder factory, Fred punched in at the time clock and worked four hours until lunch at noon. He got thirty minutes to run outside to buy a sandwich from a food truck and then gulp it down. At twelve-thirty he went back to work for four more hours. He worked hard, because the foreman often threatened to send him back to Fernald. For the same reason, he also agreed to work Saturdays whenever he was asked.

In the evening, Fred rode a bus back to Waltham. He ate dinner in a coffee shop or a pizzeria. Ordering was easy in those kinds of places, because he wouldn't have to read a menu. He would just look at the food on the plates that passed by, and ask for what he liked. Sometimes he would see a movie after dinner, but usually he would just go home to his room and listen to the radio before falling asleep. He didn't spend his quiet moments thinking about the hard circumstances of his childhood. He didn't dwell on the loneliness, loss of an education, abuse, and neglect. For one thing, all the people he had known at Fernald had similar experiences, so he didn't think that his life was remarkable. For another, he was more concerned about making the most of his future in a confusing world.

The hardest part about Fred's life during those first months of freedom was the monotony. He didn't enjoy working on an assembly line, and he found the little town of Waltham to be a bore. Looking around, at the people he saw on the street and the men and women who worked at Griffith Ladder, he felt a sense of dread. He didn't want a lifetime job making ladders, no matter how good he was at it. And he did not want to limit his social contacts and home life to Waltham and a little room on Harris Street. The world had much more to offer, and he was building up the courage to go get it.[16]

Many State Boys were released in the spring and summer of 1960, including Billy Mason, Willie Adams, and Earl Badgett. School staff reported that psychotherapy, previously deemed useless with the feebleminded, was actually helping to prepare them. In the stiff language of the profession, a psychology department report noted, "Although changes have been moderate, it is felt that this adjustment device is of value."[17]

Counseling, in groups and one-on-one sessions, was gaining popularity at many institutions. The same was true for education, as more attention was paid to preparing brighter boys and girls for life on their own. Education, psychology, and social work were coming to dominate the field of mental retardation, replacing medicine. By 1960, ten years had passed since educator Paul Hungerford had become the first nonphysician to edit the influential *American Journal of Mental Retardation*. No medical doctor would ever again hold that influential post. Hungerford and his successors would slowly advance the notion of moving as many children as possible out of institutions.[18]

At Fernald, in the top-floor wards of the Boys Home, those who were not yet paroled grew ever more impatient. One week after his best friend Fred had begun work at Griffith Ladder, Joey Almeida became depressed. His social worker noted that Joey was in a sullen funk when he arrived for his weekly counseling session. The boy explained that the Fernald girl he had designated as his sweetheart—Sandy—had been paroled. He missed her.

Joey had begun to develop an elaborate fantasy about going to live in his father's house at 10 Pinehurst Avenue in Billerica. In a couple of visits to the place, he had noticed that it was crowded with stepbrothers and stepsisters, but he liked it. He was also beginning to look forward to establishing a relationship with his father. Joey had always been impressed by his father's World War II service in France. He also liked the way his father looked. He was tall and slim, with a dark and weathered face.

The idealized father who grew in Joey's imagination did not suffer from any of the elder Almeida's obvious flaws. In reality, Joey's father had a serious problem with both alcohol and personal responsibilities. Earlier in the year, his father had called on a Friday and promised the next day to bring Joey home for an overnight visit. The boy arose early, got washed and dressed, and waited. His father never came.

As unreliable as his father and stepmother were, Joey believed them when they promised to take him in when he was released. (In Joey's mind, the seriousness of this offer was confirmed by their request that he pay $20 per month rent. If they were going to get something out of the deal, he thought, they would surely follow through.) In May, he spent a weekend there and liked it so much that he began to resent even more that he would have to wait to be free.

In June, Joey began to avoid going to his job at the campus bakery and skipped his counseling sessions. Early on the morning of June 17, he and another State Boy broke into the Fernald storeroom and took some food and clothes. When the other boy turned Joey in for the crime, Joey cornered him on the top floor of the Boys Home and beat him. The attendants who broke up the fight brought Joey to the Administration Building for a meeting with Malcolm Farrell. Those few minutes with the superintendent would remain a vivid memory in Joey's mind for more than forty years.

Farrell reminded Joey that he was scheduled for parole in a few months. Joey said that he wasn't sure that he was willing to wait.

With that, Farrell postponed his parole for an additional three months.

Joey said little as the superintendent lectured him about how he must cooperate in order to earn his release. He thought about how Farrell had named him as one of the good boys he wanted returned from Bridgewater. Then, once again, Farrell went easy on him, sending him back to the Boys Home instead of throwing him into the more secure Ward 22.

At 10 P.M. that night, Joey Almeida threw his few worldly possessions—a few items of clothing, cigarettes, and toiletries—into a paper sack and slipped out of a window at the Boys Home. He walked to Waverley Square, where he caught a bus to Harvard Square. From there, he took another bus to Arlington Center and then hitched a ride up state Route 3A to Billerica. He reached his father's house after midnight and awoke his stepmother by pounding on the door. When she finally opened it, he announced, "I'm home, for good."

Freedom was much less predictable than life in Fernald. Within days, Joey was hearing about his responsibility for rent. His father pushed him to find a job at a bakery. It was slow going, so he took temporary work as a delivery boy for a millinery wholesaler in downtown Boston. He carried orders for cloth, buttons, zippers, and thread to dressmakers and tailors.

Three weeks into his new life, Joey appeared at the back door of a bakery in Somerville, a small city just across the Charles River from Boston. Inside, the boss questioned him about his age. Joey swore he was eighteen. He also insisted he could get from Billerica to Somerville by five o'clock every morning. Using his thumb, and local buses, Joey got to work on time the following morning and every morning after that.

Thirty days passed before a Fernald social worker finally knocked on the door at 10 Pinehurst Avenue. She came not to capture Joey and return him to the school, but to deliver the news that he was indeed released from state custody. She wanted him to come back to Fernald for a meeting with Dr. Farrell, so he could be properly paroled and supervised. Joey refused. "I didn't commit any crimes," he said. "I wasn't convicted of anything, and I'm not going to accept that I'm on parole."

The woman from Fernald tried to make it sound as if Joey were declining to receive some important benefit associated with parole. She wrote in her report that because of Joey's "bad attitude, he could not be offered other services of the school." For Joey, who thought of Fernald as the prison where he had been unjustly isolated, assaulted, and humiliated, discharge was an unqualified victory. Unlike those who had gone along with Fernald's program, he had beaten the system. He would not be on parole. No one would be looking over his shoulder. More than all the other State Boys, he was free.

For the better part of a year, Joey felt sure that the job in Somerville and the house in Billerica would anchor his new life. At the bakery, he was given increasing responsibilities and even improvised a new way to make Danish so that the jelly wouldn't fall off. This earned him a raise.

The time Joey spent on his own was more challenging than the time he spent at work. On his day off, he would go to Boston, where he would meet some of the other recently released State Boys. They hung out at a cafeteria called Hayes-Bickford and at the White Tower hamburger joint across from Symphony Hall. (Part of a national chain, the White Tower was a tiny little place with medieval architecture including false parapets. It sold burgers for a dime, coffee, soft drinks, and pie. The best thing about it was that it never closed.) The boys played pool at a place called The Mines, visited Skippy White's record shop, and went to the movies to see hits like *Butterfield 8* and *Psycho*. If they had money, they might bet on a horse at a coffee shop and bookie joint called the Busy Bee, or they might listen to music at one of the many nightclubs in the center city. At some point, Joey went into a place called Joe Bett's and bought himself a switchblade, which would get him in trouble back in Billerica.

The incident began when a couple of toughs who had teased Joey every time they saw him spotted him in front of a certain pizza shop in Billerica. They stopped their car, got out, shouted that he was a "spic" and told him to go back to Boston and his "nigger friends." The taunts led to some pushing and shoving. In a matter of seconds, Joey had cut one of the boys behind the ear and the other in the chest. He fled before the police arrived, but they tracked him to his father's house. A court-appointed lawyer got Joey out of assault charges, convincing the judge that the knifings had been an unintended outcome of the scuffle.

At home, Joey's stepmother began to see him as more trouble than $20 per month was worth. She raised the rent. Then came the night when she called the motel where Joey's father was staying while working on a fence-building job in Maine. A woman answered. The battle that then raged between Joey's stepmother and father would last for months. Joey became convinced that he should move out. On a Saturday in the summer of 1961, he packed two suitcases and appeared in the kitchen of his father's house. Anthony Almeida was sitting at the table. When he saw the bags, he told his son that if he left he would never be welcomed back.

As he opened the door to leave, Joey promised to never again ask for his family's help, and he never would. When the door closed behind him, he walked about a half mile to a bus stop, where he caught the first of two buses he would need to reach Cambridge. There he transferred to the Boston subway. He was headed downtown.[19]

Charles Darwin's cousin Sir Francis Galton coined the term "eugenics" and was the leading proponent of racial improvement in the years 1860 to 1890. A polymath, he explored the tropics, did research in psychology and metrology, and invented the process of fingerprint identification. However, these achievements have been overshadowed by the racist implications of his writings on heredity and national identity.

LA SALLE NOW
Jones, Linick & Schaefer's
MADISON ST. NEAR CLARK
The World FAMOUS Chicago Surgeon
Dr. Harry J. Haiselden
in THE
Black Stork
By
JACK LAIT
A vivid pictorial drama that
tells you why Dr. Haiselden is
opposed to operating to save
the lives of defective babies.
9 A.M. to 11 P.M. | All Seats 25c
Continuous
(NO CHILDREN ADMITTED)

Released in 1917, the movie The Black Stork *told the story of Chicago surgeon Harry J. Haiselden's eugenics-based campaign to deny life-saving care to babies born with congenital anomalies. "There are times when saving a life is a greater crime than taking one," said Haiselden. Though he was investigated by prosecutors, Haiselden was never charged with a crime. He was widely supported among Progressives, who revered the seemingly objective judgments of scientists and physicians.*

Lewis Terman adapted Alfred Binet's intelligence test for use in America. One of his important innovations was the system for converting scores to mathematical IQ, or intelligence quotient. Terman was most interested in promoting the development of intellectually gifted young people. Late in his life he regretted the use of IQ tests to classify and institutionalize children.

Some people are born to be a burden on the rest.

Eugenics organizations tried to appeal to the public's sense of duty and community in its efforts to improve the national gene pool. This display, presented at the Kansas State Fair some time in the years 1925 to 1930, implores visitors to do their part to ease the burden imposed by the disabled.

The ultimate expression of "positive eugenics" was the ideal American family, which produced strong, healthy, intelligent children. Such families competed for honors in "Fitter Family" competitions. The winners of the "small family class" at the 1925 Eastern States Exposition in West Springfield, Massachusetts, are shown in this photo.

5

6

Between 1900 and 1940 local and state eugenics organizations conducted massive public relations campaigns intended to promote big families for those deemed healthy and desirable while advocating restrictions on childbearing for those regarded as "unfit." Public exhibits were mounted at fairs and conventions to promote both positive and negative eugenics. Pictured here is the "Exhibit and Examination Building" for a "fitter families contest" at the Kansas State Fair. This photo is from the American Eugenics Society scrapbook, 1929.

7

Funding and political support for national eugenics organizations, especially the Eugenics Records Office of Cold Spring Harbor, New York, was provided mainly by prominent families of industrialists and bankers including the Harrimans and the Morgans. They would surely have approved of the pro-eugenics public demonstration that took place on Wall Street in 1915. The demonstrators were hired to carry signs asking provocative questions such as, "I cannot read this sign. By what right have I children?" One of the men said he had been hired to participate in the sidewalk display by the Medical Review of Reviews, an organization that conducted contests to find the "perfect eugenic marriage."

8

As superintendent from 1887 to 1924, Walter E. Fernald more than tripled the size of the Massachusetts School for the Feeble-minded. He was widely published, enjoyed a national reputation among eugenics crusaders, and exerted such a powerful influence on the school that, after his death, it was named for him. Late in life, Fernald softened his views on the limited potential of so-called morons and even began to argue against their long-term incarceration. It was too late, however, for him to halt or even slow the rapid growth of this practice, which he had done so much to encourage. Fernald died in 1924 at age 65.

This aerial photograph shows the campus of the Walter E. Fernald State School sometime in the 1930s. The building at the center, with the white cupola, is the administration building. It is flanked on the right by the auditorium. In the foreground, behind the auditorium, is the building that contained Ward 22, the punishment facility. The cluster of buildings in the background includes dormitories and staff housing. All of the cultivated acreage pictured, along with the baseball diamond at left, belonged to the school.

The rag rugs made by both male and female residents of Fernald were distributed to state institutions for use in wards and offices. This undated photo shows preadolescent boys in the Manual Training shop of the Fernald School. They are working with the strips of fabric that hang from baskets placed above their heads.

From the Fernald School's earliest days, so-called manual training was used to teach residents basic work skills. The shops were also expected to produce useful items for the school and other state agenci. Although this photo was not dated, it shows the woodworking shop as it was well into the 1950s. Th boy in the foreground is installing bristles in a broom. Around him boys assemble benches and stools. The straw brooms they manufactured were used by boys who cleaned the Fernald School's many bui ings. They were also wielded by some staff members who used broomsticks to administer discipline.

At the Fernald shoe repair shop, boys fixed stitching and resoled boots and shoes for thousands of residents. State Boys who received no support from their families especially resented having to repair the uncomfortable work boots—they called them claw hoppers—that they were required to wear.

In this photo from around 1950, recently hired gym teacher Rose Terry leads a group of Fernald boys in calisthenics. Terry, who would later recall that she wept over the living conditions at the school, was responsible for teaching more than 1,200 students, a task she called "impossible."

Textile shops at the Fernald School produced thousands of rugs, towels, blankets, and other items for use at the school and in other state-run institutions. The young women here are working on Todd looms, which were powered by pedals.

15

Psychologist Catherine Chipman administered tests to many of the children who entered the Fernald School in the 1940s and 1950s. Doris (Gagne) Perugini would recall being assessed by Chipman on her arrival at the school and feeling so anxious and bewildered that she struggled with even the simplest tasks. The tests that Chipman used, and the methods employed to administer and score them, were flawed and unreliable by today's standards.

16

Doris (Gagne) Perugini was a teenager when she was removed from foster care and placed at the Walter E. Fernald State School. There she was reunited with her brothers Albert and Robert who had already spent years at the school. In this photo from 1954, Doris, then eighteen years old, stands on a leaf-littered lawn in front of the Fernald Administration Building.

Albert Gagne is dressed in state-supplied "Sunday clothes" for this photo taken in 1954 on the lawn of the Fernald Administration Building. Although he is quite small, due to malnutrition in early childhood, he was fifteen years old when this photo was taken.

The difference in size caused by Albert Gagne's inadequate diet in early childhood is clear in this photo, which shows him at left with his brother Robert. Albert is about fifteen years old in this picture. Robert is about age fourteen.

18

The Gagne brothers, Robert on the left and Albert on the right, stand outside the gymnasium at the Walter E. Fernald State School. They are in their mid-teens. In a few years, Albert would become the first boy allowed to leave the school under a new program intended to ease school residents into life on the outside.

20

Occasional dances gave the residents of the Fernald State School an opportunity to dress up in clothes donated by civic groups and gather in the gymnasium. In this photo from the mid-1950s, ten young men from the Boys Home sit on benches. Fred Boyce appears on the far left in a dark suit with the cuff of his white shirt showing. Seated in the front row, on the left, is Charles Dyer. Dyer would one day testify before a Congressional inquiry into the Science Club experiment.

WALTER E. FERNALD STATE SCHOOL
NAME OF INSTITUTION

CASE RECORD FOLDER

Boyce, Frederick   ADMISSION NO. 10,130

April 25, 1949

Custodial   If custodial, committed by Suffolk Probate Court.

Years in United States.   8-3

January 12, 1941   Place of Birth   Boston, Mass.

Fred Boyce   Place of birth of father. Boston, Mass.

Nina Oliver   Place of birth of mother. Boston, Mass.

Religion   Sex   Clothing
Catholic   Male   Free

: Neg.5/5/49

Congenital

Acquired .

Grade reached in school before admission. Never attended.

Mental Age.   4/4/52)8.8---77   10/25/55) 9.10--69
                   3/31/53)9.1---74   5/10/56)11.2---77
Intelligence quotient. 4/8/54)10-6---80   4/18/57)11.0--79
3/27/49)8.2---73
5/21/50)7.4---78 11/12/54)See Psycho- 3/11/58) 12-2---81
5/21/51)8.2---79   logical reports.   6/59) WAIS IQ 91

Provisional Diagnosis. 1. Moron, 2. Social Problem, Institutional, Probable,
                       3. Germ Plasm.
Final Diagnosis. Dull Normal.

Clinical Diagnosis   Familial - Moron.

Discharge, date of   March 29, 1961

How discharged.   Parole Confirmed.
(specify "from visit," "parole," "not returned," "escape," etc.)

Capability on discharge   CSS

Death, date of.

Cause of death.

        Primary

        Contributory.

Autopsy   Yes   No

Place of burial

Case files, which were maintained for every Fernald School resident, began with a cover sheet that was filled out on the day of admission. Typical was Fred Boyce's cover sheet, which reported his "clinical diagnosis" as "familial - moron." This diagnosis depended on his early IQ score of 69, which is also noted on the sheet. The cause was deemed to be hereditary or "familial" because Fred's mother had been a state ward. His experience in multiple foster care homes, beginning at the age of one, was not taken into account. Given Fred's minimal education, his final score of 91 is most likely a reflection of an improved test and better testing procedures.

22

Soon after his release from the Fernald School, Fred Boyce (left) became friends with Stewart Aucoin (right). Co-workers at a ladder manufacturing company in Waltham, they quit when they discovered they were being paid less per hour than others. They moved to Boston and rented an apartment together in the Back Bay neighborhood.

23

Albert Gagne met Doris Wheeler in the early 1960s at a shoe factory where they both worked in Lewiston Maine. They would soon marry and have one child, a daughter named Karen. Eventually they settled into a home on wooded acreage that is part of the Wheeler family farm

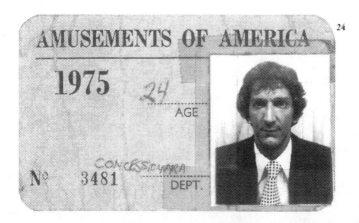

After gaining experience working with his friend Robert Catalano, Fred Boyce bought his own game of chance and took to the traveling carnival circuit. By 1975 he was a concessionaire for a carnival operator called Amusements of America. His listed age on this identification card—24—is ten years younger than his actual age.

Abra Glenn-Allen married Fred Boyce in 1987, at about the time this photo was taken in the living room of Fred's house. She entered the Peace Corps a few months later and served overseas for one year. The marriage would not survive this separation, and they would eventually be divorced. However, Abra would remain Fred's lifelong friend and help care for him when he was treated for cancer in 2002.

Once their lawsuits against the Commonwealth of Massachusetts, MIT, and Quaker Oats were publicized, the reunited members of the Science Club became momentary celebrities. This photo appeared in People *magazine, which described their suit as a "Revolt of the Innocents." From left are former Fernald State Boys Gordon Shattuck, Larry Nutt, Lester Foye, Fred Boyce, Austin LaRoque, Charles Dyer, and Joseph Almeida.*

Despite a painful separation and divorce, Fred Boyce would remain close to his former wife Abra and her family. He is pictured here in 2002 with Abra's daughter, Olivia, and her son, Julian, after the successful settlement of the Science Club lawsuits and his recovery from surgery and chemotherapy for cancer.

This photo shows the Boys Dormitory, once home to young males at the Walter E. Fernald State School, in 2003. Long abandoned, the building is fenced to keep out vandals as it deteriorates. The Commonwealth of Massachusetts planned to close all of Fernald in May 2004 and some state officials expect the property to be sold. A buyer would acquire more than 100 acres of land and buildings in various states of disrepair. The site's location close to Boston and the scarcity of open land in the area would make it quite valuable.

NINE

Eighteen-year-old Joey Almeida, short and stocky with his curly black hair cut close, emerged from the Boston subway at the Park Street station with his suitcases and no firm idea about where he was going. He hoped to run into some of his old friends from Fernald. Failing that, he thought, he would check into a cheap hotel and wait until dark, when he could find them at the White Tower.

The wandering approach to life was not unusual among the State Boys who had been released from Fernald under the new program. Although the school's social workers had found rooms in the suburbs around Boston for each of them to occupy upon their parole, boredom and curiosity led many to abandon these places for the big city. There they often found each other and shared apartments until one or two rent payments were missed and they had to move again. (Exceptions to this lifestyle included the Gagne brothers, who preferred the safety of small towns.)

The Back Bay neighborhood, where Joey went to find his friends, was an odd mix of high and low. Symphony Hall and the nearby Horticultural Hall exposition center attracted the wealthiest Bostonians. But for blocks in every direction, the streets were lined with cheap restaurants, pool halls, and crowded apartments.

Rents were low in the Back Bay because the middle class was fleeing the city for the suburbs. Boston had lost 15 percent of its population in the 1950s. Business suffered, and empty storefronts blighted many streets. To revive the urban economy, city officials began massive "urban renewal" projects in an area called the West End and at infamous

Scollay Square, a fading honky-tonk collection of bars, nightclubs, and the remnants of burlesque. Over the decade to come, new housing and an office complex called Government Center would rise on these sites.

The renewal projects had displaced thousands of people, many of whom went to the Back Bay. It became a lively, if somewhat dangerous, inner-city haven for the working poor, recent immigrants, college students, pensioners, and hustlers. Young men like Joey, who had little money and an appetite for excitement, loved the place.[1]

At least a dozen former residents of the Boys Home lived in the Back Bay. They "mobbed-in together" in small apartments, splitting rents of $50 or $60 per month. A few of them, most notably Billy Mason and Earl Badgett, joined a local gang called the Majestics and could have been called low-level hoods. But most were working menial jobs and just flirting with trouble. At the White Tower and other all-night spots, they found acceptance among those who lived just outside of proper society.

When he got to the Back Bay, Joey hadn't walked more than two blocks when he heard someone call his name. It was Fred Boyce, riding in the passenger side of an old Ford. The car pulled to the curb and Fred introduced Joey to the driver, his new friend Stewart Aucoin. In less than a minute, Fred and Stewart invited Joey to stay in their apartment at 57 Westland Avenue. Joey put his suitcases in the car, and they took him to their place.

In the six months since Fred had been paroled, he had found that life on the outside could be as full and colorful as Fernald had been empty and gray. At the ladder company, he and Stewart had figured out that they were being underpaid compared with other workers. Since they spent all of their off time in downtown Boston, they decided to quit, leave Waltham, and move down there.

At first, Fred and Stewart found shelter with a couple of older parolees from Fernald who were just beginning a lifetime together as gay partners. After a week or so, the apartment began to feel overcrowded and the expenses of feeding the two guests, who had yet to find jobs, began to be a burden. Sensing this, and noting that one of his hosts was about to celebrate a birthday, Fred thought up a scheme to solve the food problem and give a generous present with a single risky act.

Late one night, Stewart and Fred borrowed a car and drove to Waltham and entered the Fernald campus via a little-used service road. They parked the car near the warehouse, and Stewart waited while Fred crawled in through a window. Once inside, he collected boxes of canned goods and bags of coffee, hoisted them up to the window, and let them fall to the ground below. After loading the car, Stewart and Fred returned to Boston and presented their loot as gifts. Their hosts were so appalled by what they had done, and frightened that they would be caught, that they asked them to find jobs so they could move out as soon as possible.

Desperate to get some income, Fred briefly worked for a company that paid him to show up at Fenway Park at the end of ball games to clean the stadium. He spent hours working from row to row, tilting up seats so that he could sweep up paper cups, candy wrappers, spilled popcorn, and peanut shells. Though it was one of the smallest in the big leagues, Fenway seemed like a huge stadium when you were supposed to pick up every scrap of litter that the crowd left behind. Fred hated the tedium, and the mess. Baseball fans could be incredible slobs. Hoping for something better, Fred applied for a job washing dishes at the lunch counter at the Hemenway Drug Store, which occupied a busy street corner in the Back Bay. He got the job, and soon he and Stewart found the apartment on Westland Avenue, just a few doors away from the store.

Fred's neighbors included a group of professional wrestlers—Haystacks Calhoun, Golden Boy Dupre, the Mad Russian—whose apartment was their home base when they weren't on the road. The mere presence of these massive men meant that even though break-ins and fights were common in the neighborhood, their building was almost completely crime-free. But the wrestlers didn't intimidate everyone. Kids used to taunt Calhoun on the street, knowing that he was reluctant to take them on for fear of hurting them. He took to wearing a heavy chain around his waist and swinging it at his tormentors so that he could shoo them away without actually touching them.

The Back Bay was filled with students who attended nearby North-eastern and Boston Universities. On Friday and Saturday nights, Fred and Joey would dress in jackets and ties and walk the neighborhood until they found a party in a student apartment. Whether they knew the host or not, they would go in and mingle, drink, and smoke. They had both taken up cigarettes. Fred smoked two or more packs a day. But they avoided the drugs that could be had at the parties, or under the counter at a drugstore called Ruthie's. Cocaine was not yet popular, but the store did a brisk business in codeine cough syrup, amphetamines, and sedatives such as Seconal and Tuinal.

When they couldn't find a party to go to, Fred and his friends made their own. Quarters, dimes, and the occasional dollar were pooled to buy cheap beer and snacks. Anyone who appeared at the door was welcomed. This policy often led to big crowds and complaints from neighbors. Not long after Joey had appeared in the Back Bay, Fred came home from work one night to find his apartment overflowing with people. The noise eventually drew the Boston police, who pounded on the door.

The police in the local precinct were especially aggressive, and they knew many of Fred's crowd on sight. Like other major cities, Boston was struggling with what was perceived to be a juvenile crime wave. The city became especially alarmed in the summer of 1960 when a gang jumped the sons of both the president and the dean of Tufts University as they came out of a drugstore. Local charities began sending teams of social workers into neighborhoods to offer street counseling to an estimated 250 gang members. But for the most part, the city depended on police to control gangs, and they did the job with informants, threats, and a measure of physical force.[2]

When the police came to Fred's apartment and he opened the door for them, the guests began to jump out the window and climb down a fire escape. Some also slipped past the officers and down the hall. With so many people fleeing, the officers began to detain whomever they could grab. Fred and more than fifty others were handcuffed and walked out to the sidewalk, where they were eventually loaded into paddy wagons and taken to a precinct house. It was there that Fred learned that the police had found drugs, a pile of switchblades, and a handgun in the apartment. Since his name was on the lease, the officers assumed that it all belonged to him.

At the station house, Fred continued to insist that the party had been going on when he arrived home from work. After an hour of fruitless questioning, the police officers asked him where he worked. When he told them that he bused dishes for the lunch counter at the Hemenway Drug Store, their demeanor changed, and Fred began to understand more about how life worked in the Back Bay. It turned out that the drugstore's owner, Nick Elias, was well connected to the police. One call and Elias was at the station. Fred was released into his custody. Elias also arranged for a lawyer, who cleared Fred of all the charges associated with the party and the police roundup at his apartment.

Nick Elias would turn out to be a strong, stabilizing influence on young Fred. He taught him to be punctual, speak politely, and work hard. In a short time, Elias promoted him from the dish room to the counter. Always talkative, Fred found it easy to chat with the customers, who included a number of single, middle-aged women who were charmed by his good looks and genuine interest. A few of these women, nurses who worked odd shifts, became regulars who stopped in for coffee and a chat on their way to or from work.[3]

Hemenway Drug closed at nine o'clock every night. After that, Fred, Joey, and their buddies might go to a movie, or a restaurant, or play pool at The Mines, which was on Washington Street. The Mines was full of hustlers, and they soon learned to say that they didn't have any money in their pockets. That was the best way to avoid making bets that they were sure to lose.

Almost every night out ended at the White Tower across from Symphony Hall. It was a perfect hangout because it never closed, the menu was full of 10-cent items, and the clientele was endlessly fascinating. Singers, comics, and entire bands that played at local clubs came looking for after-hours food. Cabbies stopped for coffee and told stories about their fares. Prostitutes and pimps trolled for johns. In one White Tower escapade that Fred and Joey would both remember for decades, Joey decided to play traffic cop outside Symphony Hall just as a performance ended.

It all began with Stewart Aucoin's hat. He had gotten a job driving for Checker Cab, and the company required him to wear a hat that looked just like the short-billed hats worn by Boston police officers. Joey asked to borrow it, along with a whistle that Stewart kept in his car. As the concert ended and the crowd came out of the hall, taxis, chauffeured limousines, and private cars began lining up on the street. To the delight of everyone inside the White Tower, and to the torment of every driver, Joey began directing traffic. He whistled and barked with the gruff authority of a public servant who enjoyed pushing the tuxedo crowd around.

Soon Joey had cars jammed together at ninety-degree angles, and drivers leaning on their horns. A few of the symphonygoers who stood on the sidewalk began to understand what was going on and laughed. The drivers who were gridlocked didn't think it was so funny. Joey, Fred, and Stewart fled in Stewart's cab, which was parked away from the jam-up, before the real police arrived to straighten things out.

Though he appreciated the brazenness displayed by his friend Joey, Fred expressed himself in quieter ways. He liked fashionable clothes, and he dressed so well that his White Tower nickname became "Suit Jacket." His maturity impressed the manager, who offered him a job. Fred thought that counter work was hardly work at all, and he liked the idea of getting paid to come to a place that was already a hangout. He was happy to put on an apron and a white paper hat and take responsibility for the overnight shift.

Fred loved meeting customers and chatting with them about the weather, the news in the paper, or the Red Sox. He believed that each person he served might teach him something new about the world. And as long as the menu was simple—burgers, drinks, and slices of pie—he had no trouble filling orders quickly and keeping folks satisfied. Fred even liked the quiet that came between 3 A.M. and dawn, when he was often as alone as a lighthouse keeper. Eager to develop his vocabulary, he bought a dictionary, which he set on the counter and studied during his breaks.

Late hours did not always guarantee peace and quiet, especially on weekends. Many of the customers who came into the White Tower during these hours were drunk or drugged and acted in unpredictable ways. One morning, a man in an expensive suit who flashed an equally expensive watch sat down and began talking about how Fred wasn't fast enough or friendly enough. When Fred made a remark about how the man was a little drunk, and a little belligerent, his customer's reply came as a surprise.

The well-dressed man said he was a parole officer and that he just knew that Fred was on parole. Moreover, he said that Fred must surely be violating the terms of his release, and he might just do something about it.

Suddenly reminded of the way the state loomed over his life, Fred said he had never been in jail and that he doubted the man's claim that he was a parole officer. Then he made sure the man knew that he was a tough guy. "I've got a black belt in karate," he lied. "And I can do a flip over this counter without using my hands."

Charmed by this turn in the conversation, the man in the suit smiled even more broadly as Fred agreed to bet $5 that he could do the trick. He knew the boy was bluffing, and let his jacket fall open just enough to reveal the pistol he carried to enforce wagers. Fred looked at the gun, smiled sheepishly, and just handed him a $5 bill.

With their roles established—the counter boy and the tough guy—Fred's customer confided that he was a professional gambler, "the kind that never loses." Fred knew this was impossible. Even the best players at The Mines lost once in a while.

"No," replied the man on the stool. "I always win because once everybody's relaxed, I rob 'em."

The gun. The man's clothes. The attitude. They all persuaded Fred that his lonely customer was telling the truth. He tried to be cool. "You could get killed someday doing that," he said. The man took the last bite of his hamburger, said something about how every businessman accepts the risk of his trade, and left.

The street characters and mysterious strangers who drifted through the White Tower gave Fred stories to tell and made him eager to come to work. But he wasn't always able to avoid being part of the scams and crimes that bloomed in the middle of the night. Before dawn one morning, a young tough he knew from the Back Bay came in demanding that Fred turn his back so he could break open the jukebox and steal the coins inside. He needed the money to pay his lawyer, and he badgered Fred until he relented. When the boss arrived in the morning, Fred lied and said that someone had stolen the money in a moment when he had gone into a back room for supplies. He felt guilty about lying but knew the truth would cost him his job, so he stuck to the story. The boss, who considered Fred almost uniquely reliable, readily believed him.

Fortunately, most of the late-night customers Fred saw were regulars who kept him company when business was slow. They might buy a hamburg—at that time in blue-collar Boston few called them hamburg-*ers*—and then Fred would keep pouring free coffee. Many of these reliable customers were older black men from the nearby Roxbury neighborhood. In the daytime, police ran them out of the area around the White Tower, enforcing unwritten rules of segregation. At night, the rules were looser.

Fred saw the way that racism affected his black customers and connected it with his own experience at Fernald. "There was nothing wrong with these guys," he would say years later. "They were strong, grown men who worked hard but had nothing. They were just black. That was the whole thing."

A Roxbury regular came to Fred's rescue on his worst night at the White Tower. The trouble started when a drunken prostitute took one of the stools at the counter and called out her order. Fred's back was turned, and when he didn't acknowledge her, she picked up a heavy sugar bowl and hurled it at him. The bowl barely missed his head, making a dent in the white sheet metal of the wall.

Fred turned around and jumped the counter. He grabbed the woman by the arm, yanked her to her feet, and rushed her out the door. He then stood, with his foot against the bottom of the door while she put her shoulder against it and screamed about killing him.

As Fred held the door, a big car pulled up and two men got out. Fred recognized one as a pimp. He knew they were armed and that they would demand to know what was going on. Fred let go of the door, dashed back around the counter, and grabbed two pots full of hot coffee. When the pimp came in with his partner and the angry woman, Fred brandished the pots like they were a pair of six-shooters.

The sight of this skinny young man in the paper hat holding coffeepots like they were weapons made the pimp stop long enough to smile before he demanded to know what was going on. Fred tried to explain, and then a black man—a regular at the end of the counter—spoke up. He described the incident and how Fred had responded without really hurting the sugar bowl thrower.

With these two men discussing what had occurred, Fred had time to think of a way to make something good come out of the confrontation. He thought of the one Boston pimp he knew well, a man named Lance, who seemed to have dozens of women working for him. In the language of the day, he was "real strong," meaning he was highly respected by his peers. But Lance was always worried about the police. Years later, Fred recalled what he said.

"I told them, 'The next time Lance comes in, I'm going to tell him that none of his girls can come in here anymore. I can't have girls come in here and throw sugar bowls around. Any gal that comes in here, I'm going to call the cops now. And it's all because of her and you guys.'"

As the White Tower was one of the few places that never closed, every pimp and prostitute in Boston depended on it. Fred's threat, combined with the testimony of the man at the counter, was persuasive. The pimp and his sideman took the woman outside by their car and made a show of slapping her. From that night on, Fred enjoyed unbroken peace with the professionals who continued to use the shop as a place to make contacts.

The peace that Fred established on the night shift was a real accomplishment. White Tower managers recognized his success with customers and his work ethic. A district supervisor tested him by assigning him to other shops and discovered that he adapted immediately to new coworkers and new surroundings. He then called Fred into his office to discuss a promotion. Fred could be a night manager and, in time, move even higher in the company. The offer was flattering, and the extra money wouldn't hurt. And in the excitement of the conversation, Fred almost forgot that he couldn't read or write. Then the boss handed him some forms he would need to complete for the main office. Fred couldn't make out any of the questions, let alone write the answers.

"Look, I was on the state as a kid and I never really got any schooling," said Fred.

"You can't fill that out?"

"I'm sorry."

The job would have required that Fred handle paperwork for orders and deliveries. He would have had to read memos and write occasional notes. The district manager was obviously disappointed. He explained to Fred that he couldn't follow through on the promotion. All Fred would recall his saying as he withdrew the offer was "I'm sorry, too, Fred."

The lost promotion marked one of the few moments when Fred was confronted directly by the Fernald legacy. All the State Boys would have similar experiences. They either lost opportunities or got scammed because they could not read or write. More than one of them would fall victim to used-car salesmen who roped them into extended, high-interest-rate deals on bad cars. They also suffered socially because they had trouble understanding the language and rules of behavior that governed life on the outside.

"At Fernald, no one corrected you if you used the wrong word or talked funny," noted Fred. "Everyone talked that way. You'd say something like, 'That's "in-ca-nif-i-cent,"' and they knew you meant 'insignificant' and just accept it. They wouldn't even correct you. But people on the outside would notice that there was something strange."

Many former State Boys also brought the attitudes from life in the wards into the outside world. They were easily influenced by people who seemed older, smarter, or tougher, and they had trouble recognizing that they were being manipulated. On the job, the result was that they ended up with the lowest pay, bad hours, and the worst assignments. In social situations, they might send the wrong signals. Fred, for one, was very friendly, very open to people, and always smiling. More than one gay man misread this as a sign of romantic interest.

Romantic confusion was not part of the disturbing sexual encounter that occurred in Fred's second year as a free man. One evening at a coffee shop, a friend from Fernald—the Big Shot Willie Adams, who had made Joey his "boy" at the BH—introduced Fred to his new girlfriend and to an enormous man who had become a new buddy. The four went back to the young woman's apartment. Soon Fred was alone on the sofa in the living room with the giant man on top of him, trying to force him to have sex. Fred fought back as the other man held him down, pulled on his trousers, and undid his own pants. Fred managed to shout loud enough to bring his Fernald friend into the room and end it, but he was traumatized by the assault.

Many years later, Fred would recognize that he had missed signals in the conversation that should have told him to avoid Willie and his big friend. The man had been a little too interested in him and a little too eager to find some privacy. Fred and the others from Fernald often missed the rhythm in conversations and failed to pick up social cues. This was especially troublesome in certain jobs, and many Fernald parolees looked for work where they wouldn't have to talk much. It was a little like life in the Boys Dorm, when McGinn demanded silence and set the boys to work on the oak floor.[4]

Although Fernald had not given the State Boys a proper education, it had been good preparation for an environment in which intuition and quick reactions had great value. Having grown up with attendants who could turn violent in a moment, the State Boys each possessed the flight instincts of a wild deer. A gesture, a slight change in tone of voice, or a shift of body language would send them out the door. The brighter boys like Fred also knew how to avert a conflict with just the right words.

Fred's main strategy for staying out of trouble involved avoiding situations where he might be alone with the tougher characters in the neighborhood. For example, there were a number of young men who hung out in alleys and isolated corners of public parks that they called "sneak spots" because police cars could not reach them. They would drink heavily, and talk loud, and sometimes get into fights. Fred avoided those places. He also refused to succumb to pressure from those who said he had to earn acceptance by fighting or committing a crime.

"There was this guy called Squeaky, who was always trying to test people," explained Fred, years later. "We were walking down Massachusetts Avenue once and he saw this car with a camera sitting on the seat. He said, 'Freddie, you ain't never done nothin' that I know about. Why don't you break into that car and take that camera?'" Fred refused to act on Squeaky's suggestion, telling him that he had "nothing to prove" to anyone in the neighborhood. He never again went out alone with Squeaky.

Cautious about sneak spots and characters like Squeaky, Fred was nevertheless eager to experience the nightlife in downtown Boston. Sometimes he went to Crusher Casey's, a bar near the White Tower. The place was owned and run by an Irishman named Steve Casey, who had won the 1938 world wrestling championship in Boston Garden. Casey was friendly with pro athletes, who drew crowds when they came. But thanks to the size of the local Irish community, Casey's place was always crowded.

If they weren't at Casey's, Fred and his friends might walk up Washington Street to an area that would become the infamous collection of strip joints and clubs known as the Combat Zone. In the early 1960s, the seedy area was being filled with businesses that were fleeing Scollay Square. The Golden Nugget and the Hillbilly Ranch offered music and overpriced drinks. Strippers were the main attraction at the Naked Eye. At the Melody Lounge, a small community of openly gay men and lesbian women welcomed anyone who wandered in, including sailors from Navy vessels that were in port.

In the daytime, Fred and some of the other paroled State Boys often went back to Washington Street to a little coffee shop called the Busy Bee. With a nickel jukebox and 10-cent coffee, it was the kind of place where someone was always begging quarters to buy a bottle of booze, and harebrained schemes for petty crimes were a subject of constant conversation. The owner took bets on the horses that ran at nearby Suffolk Downs thoroughbred park.

The betting action, which was run out of a back room, made the Busy Bee much busier than the average coffee shop. Hundreds of people came through the door every day, including a few women who caught Fred's eye. He dated several of them, but his first serious girlfriend was a seventeen-year-old named Irene Conte. Once they became close, she would come into the hamburger shop and coax him into going out after his shift. They would go to a movie or a restaurant, and then go to Fred's apartment for a little necking. Before things went too far, Fred would take Irene home to her apartment, which she shared with her mother. Living alone, the mother and daughter were especially close. And Fred would long remember how Irene worried about her mother's relationship with a boyfriend who sometimes turned violent. She even wondered aloud whether God would forgive her if she felt compelled to shoot the man in order to defend her mother.

Having become part of Fred's social crowd, Irene began spending time at the White Tower even when he was not around. She made friends and one afternoon accepted an invitation to take a car ride with one of the young men who were regulars in the shop. Within a few blocks, the driver committed some minor traffic violation that caught the attention of a police officer in a patrol car. When he heard the police siren and noticed the flashing light, the driver panicked and raced away. The high-speed chase ended in a crash. Irene was found unconscious and, after several days in the hospital, died of her injuries.

Though Fred had not pledged his love to Irene, he was devastated by her sudden death. Months would pass before he met another young woman—older and much more self-assured than Irene—named Brenda. They dated long enough for Fred to lose his virginity. The sex was satisfying, he would recall, but again he didn't feel the spark of love. This may have been due to his fond memories of Margaret Burney, the girl he had idealized at Fernald. He saw her once in the Back Bay, after she had run away from the institution. But he was afraid that if he got close to her while she was a fugitive he would risk being sent back. Fred didn't pursue her. She dated and married someone else. He would forever wonder what might have happened if he hadn't been so cautious.[5]

Despite its Brahmin image, much of Boston in the early 1960s was wild and unpoliced. Fronts for bookies existed throughout the city. For example, hundreds of bets were made in a typical day at the Schwartz Key Shop on Massachusetts Avenue. Bookies were so ingrained in the local culture that in 1960 a mob of 1,500 turned out to oppose a police raid on a betting shop in the Italian North End. Two officers were assaulted by angry residents who shouted at the police to "Go home!"[6]

As neighborhood institutions, bookie joints were considered neutral ground among the small gangs that competed for territory downtown. Fred never joined a gang in a formal way, but he fell under the protection of the Majestics because Billy Mason, from Fernald, became one of the gang's leaders. In his first few years out of Fernald, Mason had become a very visible figure in Back Bay Boston. People called him Cheyenne, after a handsome TV western character named Cheyenne Bodie. He wore a Majestics jacket almost all the time and quickly earned a reputation that made people back down from a fight.

Fred benefited from Cheyenne's reputation when one of the characters at the Busy Bee—he was called the Umbrella Man, because he always carried an umbrella—pushed him hard over the matter of 25 cents. He had already begged 50 cents from Fred but wanted more so he could buy a bottle of Wild Turkey. The argument got heated enough for the two young men to go outside.

"He started out saying, 'Don't you know who I am? I'm the Umbrella Man,'" recalled Fred.

"Someone on the sidewalk told him not to mess with me because I was Cheyenne Bodie's brother. At first, he didn't believe it, because I didn't look like Cheyenne. But they told him we grew up together, and he finally believed me. Then he showed me why he carried that umbrella. There was a lead pipe inside it. If he had hit me right with that thing, it could have killed me."

These run-ins were a part of Fred's life for as long as he stayed in the Back Bay and frequented certain places where he was bound to encounter certain people. He talked his way out of every confrontation but one. In that instance, Stewart Aucoin had been mugged on Westland Avenue. As far as he could tell, he had been jumped because he was wearing his Navy pea coat, and thugs in the neighborhood thought sailors were easy marks.

When they heard what had happened, Fred and former State Boy Earl Badgett went to look for the attackers. With nothing more than Stewart's vague description to go on, Earl and Fred wound up confronting the wrong men. In the street fight that ensued, one of the men pulled a crescent wrench out of his pocket and slashed Fred across the face with it. He fell to the sidewalk with blood gushing onto his neck and chest. The man who hit him ran.

Although Fred was upset by the sight of his own blood, he didn't want to go to a hospital because he feared that a nurse or doctor might summon the police. The police could start asking questions and, who knows, he might be sent back to Fernald. Still, his wound needed stitching. The one place he thought he could go and receive care without much of a hassle was Thom Hospital, on the Fernald grounds. Badgett drove him there. Nurses who remembered Fred believed the tale he told about falling and getting cut by the sharp edge of a metal table. They took care of him.[7]

The experience of returning to Fernald and getting attention from the nurses made Fred feel strangely ambivalent. He had hated the place, like an abused child hates the parent who beat him. But, like that child, he also felt an attachment to his past. The fact that he was treated so well when he returned in need confused him. But then he thought of McGinn and other violent attendants and of the North Building. He reflected on the moments of torture he witnessed and the hopelessness he had felt at the BD and the BH. The anger returned, and he was again grateful for his freedom.

Though he quickly learned to navigate the streets of Boston, Fred struggled with a changing sense of his own history. Soon after he began working at the Hemenway Drug Store, he looked up from the counter to see his brother George, whom he hadn't seen since that surprise visit in the gym at Fernald. George was on his own, too, and because he had gone through public high school, he had been able to get a good job with a manufacturer of electronic equipment. George had contacted their mother and asked Fred if he wanted to meet her.

Any thought that he even had a mother had long been buried in Fred's subconscious. He had come to believe that she was either dead or long gone from the city. He was shocked when George told him that she lived less than one hundred yards from Hemenway Drug, in an apartment on Burbank Street. George suggested that they go knock on her door when he was finished with work, but Fred couldn't move that fast. He suggested they do it on a Saturday morning, after he had more time to think about how he would handle it. They picked a date, a few weeks into the future. When the day came, Fred got up early, shaved with extra care, and dressed in his best clothes.

If Fred was nervous as he walked to Burbank Street, he wouldn't remember it. He would recall only climbing the stairs to the second floor of a brick apartment building and hoping that his mother would offer just a word or two of apology. "I knew how hard life could be," he said. "If she said anything, I was ready to just let it all go and give her whatever she wanted from me." When they reached the door of Mina's studio apartment, George knocked. When it opened, he said something like "Hi, Ma. This is Freddie."

A short, thin, forty-one-year-old woman with black hair and smooth dark skin, Mina Oliver (she was no longer Boyce) was dressed in her starched white nurse's uniform. She stood back for a moment, taking in every inch of the young man who stood in the doorway and waiting to see if he would react. When he didn't move, Mina stepped forward, hugged Fred around the neck, and pulled both of them inside.

The apartment was a neat but ordinary place filled with battered old furniture and decorated with cheap knickknacks. George and Fred sat at a little table that had been pushed against one wall of the kitchen. Mina asked if they would like something to eat or drink. They wanted nothing, but Mina went to her refrigerator and got a bottle of beer for herself. She lit a cigarette and then alternated puffs and swigs of beer.

In the awkward hour that they spent together, Mina told Fred that she was a practical nurse who worked at various hospitals around the city. She wasn't married, but she was dating a man who treated her well, she said. (Later, Fred would learn that this boyfriend was an active, if more or less quiet, alcoholic.)

Gradually, Mina confessed that she had given up several other children, but she wouldn't share many of the details of what had happened. She did recall that her son Joseph, the two-year-old taken by the state along with Fred, had been adopted. Given the secrecy of the process in those days, she had no other information about him. She didn't know his full name, or if he was even alive.

As the alcohol took hold, Mina became more emotional and she began a kind of speech, one that Fred would hear again, almost every time he saw her in the future.

"She said a lot of stuff about how hard she had it when she was young. 'Your mother worked so hard to take care of you,' she would say, crying. 'It wasn't my fault that the state took you kids. It wasn't my fault. I wanted to take care of you and they wouldn't let me.'"

Every member of a family will construct a story to explain the past. Each person has a different narrative, one that turns the saga into something logical from her own peculiar perspective. In Mina's version of the family legend, she claimed the role of "first victim." The Commonwealth of Massachusetts was the villain. And the loss of her many children, which took place over time, was caused by a series of random tragedies, like tornadoes, that roared beyond her control.

Fred allowed his mother to spin the tale. Her tears prevented him from saying anything about his own experience in foster care and at Fernald. He said nothing of the abuse, the fear, the loneliness and confusion. He even offered her a way out of her regret, by suggesting that "the past is the past" and that they focus on creating some sort of relationship for the future. Fred's words were enough to stop her sobbing and allowed him and George to say goodbye. It had been an uncomfortable visit, but Fred still promised to come see his mother again. He felt some strange obligation, and besides, he was curious about the family he never knew. He let her hug him again, and then left.

As Mina's two sons walked down the steps and out onto the sidewalk, Fred thought about how familiar his mother seemed to him. It could have been something in her appearance—the nose, the angle of her chin, or her complexion—that reminded him of the face he saw in the mirror. Then, as he walked to his apartment, Fred suddenly realized that he had seen her seated at the Hemenway lunch counter. She had gazed at him once, for a strangely long moment, while he was hauling dishes up and down the stairs that led to the kitchen.

Fred wouldn't see Mina at the drugstore ever again. She stopped going there after they met. But he did visit her from time to time. More often than not, she drank and then cried as she returned to the subjects of the men she had known and how the heartless state had seized her children without cause. Mina never offered Fred one word of apology for abandoning him as a baby.

Guarded when it came to the true details of her life, Mina would take years to tell Fred that the man she had named as his father, Ford Boyce, was a good man who had been ruined by his experiences in the war. Fred also caught hints that made him think that in fact he was another man's son, but he would never get her to admit this. He also learned that he had many siblings, perhaps as many as a dozen. They had all been wards of the state.

Following the clues that Mina dropped in conversations, Fred found a twelve-year-old brother named Steven who lived with a foster family. He visited him on weekends, taking him to the beach and to concerts. Once they saw Roy Orbison perform. Fred also found a sister named Donna, and a slew of cousins. As late as 1985, he would track down a second sister, Denise, who lived in the suburb of Roslindale. He was so excited on the day he drove to see her that he rounded a corner too fast and skidded into a pole by the side of the road. His van was not seriously damaged, so he just drove on. Denise greeted him with a big embrace and became the sibling who welcomed him for holiday visits every December.

Mina never laid out the family tree in any logical sense, so Fred never learned the names of all of his brothers and sisters. But many of the relatives he did meet, especially cousins, struck him as unreliable. Warm one moment, distant at another, some even turned violent at the most unexpected moments. Gradually, he concluded that even if he had escaped being sent to Fernald, he would not have had any ideal home life. "Still, it would have been better than an institution," he would say. "Think about how well I coped with Fernald and you can imagine what I would have done with a family. *Any* family at all would have been better."[8]

One year after his parole from Fernald, a social worker from the state school made one last visit to Fred Boyce. He documented his income for the year—$1,853—and noted that he had saved $150. (A respectable percentage of his earnings.) He reported that Fred said he was happier in a steady job that paid less per hour than in the kind of seasonal work offered by the cleaning company at Fenway Park. He also valued the relationships he had with the owner of Hemenway Drug, Nick Elias, and with customers and coworkers. "He receives personal consideration and recognition," wrote the social worker, "which compensates for any lack of monetary rewards."

The picture painted in the social worker's report is of a friendly, capable young man who worked six days a week and handled his money well. "He has regularly demonstrated responsible, self-dependent behavior and needs no further supervision from our department. He is ready for discharge, which he has been for some time."

On March 28, 1961, the same social worker went to a meeting in the Fernald Administration Building to present Fred's file—more than nineteen years' worth of notes, reports, and test results—and his recommendation that he be permanently released from the commonwealth's custody. The group that gathered around the superintendent's table included physicians, psychologists, Dr. Benda, and Malcolm Farrell. Fred's discharge was approved unanimously, pending his registration with the draft board in Boston. The registration papers were filed the next day, and Fred was freed. He was twenty years old and had spent all but the first eight months of his life as a State Boy.[9]

Fred was fortunate to have won the trust and confidence of the staff at Fernald. He was also lucky to have been considered for discharge in 1961. In the next year, the so-called Boston Strangler would provoke widespread fear, even panic, and Fernald's male alumni would come under intense scrutiny. This was because the strangler's eleventh victim was Beverly Samans, a twenty-six-year-old woman who worked part-time at the state school. Police focused much of their investigation on former State Boys. One, named Daniel Pennachio, confessed to the crime. Fernald doctor Adrian Blake would recall that "poor Daniel also said that he killed Jack Kennedy and Abraham Lincoln."

Daniel Pennachio died in a swimming accident shortly after his confession. Adrian Blake would never believe he had anything to do with any murders. Construction worker Albert DeSalvo eventually confessed to the murders, and the stranglings stopped. However, some skeptics have long considered DeSalvo's statements to be false, and the Fernald connection to the stranglings never faded. As late as 1995, Daniel Pennachio's name was raised again in a book that alleged that he or "someone very like him" killed Samans.[10]

By the time Fred received the news of his permanent discharge, he had begun to develop some bigger ambitions. He was outgrowing the Back

Bay lifestyle and began to fantasize about moving to a comfortable suburb, perhaps even buying a house. He knew this would take money for
a down payment, and he didn't quite know how to get it. But the
desire to have a secure place of his own was fixed in his mind. He
began looking at maps, identifying small towns within a certain radius
of downtown Boston. He figured thirty miles was a maximum distance
for anyone who might commute to work.

Fred's interests in real estate and commuting were not even considered by most of the other State Boys he still knew. Many of them were
too timid to even think about getting a driver's license and exploring
beyond their immediate neighborhoods. Fred would long remember
discussing highways with a former Fernald inmate. He had to go to
great pains to explain that once he got on Interstate 95, the highway's
name and direction wouldn't change. The man had been confused by
exit signs where route numbers were posted. He thought that each
passing sign indicated he was on a new road.

Many other State Boys would have considered Fred's longing for
convention a boring sellout. Young men from the Back Bay did not
pine for domestic bliss. They were interested in excitement. Action.
Billy "Cheyenne" Mason and another Fernald alumnus named Ronnie
Beaulieu, for example, were deeply involved with the Majestics gang.
They wore gang jackets and spent their nights drinking, carousing, and
flirting with crime. Others, like Joey, had no time to think about where
they might be at age thirty, or forty. They were too preoccupied with
chasing young women and finding their next dollar.

One of the quickest and easiest ways to get a few dollars without
working involved a heavy screwdriver and a stroll down almost any city
block. After glancing around for police and witnesses, Joey would jam
the blade of the screwdriver into the lock of a parking meter and, with
the twist of his wrist, break it open. Inside, he would find a paper cup
holding the coins that had been deposited since the most recent collection. Sometimes working with a partner, Joey might break every meter
on a block and take so many nickels and dimes that he could buy two
steak dinners at Ann's, a sit-down restaurant that was near the White
Tower.

Joey learned how to break into meters from new friends he met in the city. They also taught him the simplest way to steal a car. In the early 1960s, the ignition systems in many cars required drivers who had just parked to turn the key counterclockwise to lock out thieves. If they pulled the key out before completing this action, the system would be left in the "on" position. If the car door was unlocked, and many were, no effort was required to get in, turn the flanges on the keyhole, and start the car without a key.

At first, Joey took cars only when he needed them for immediate transportation. He would break a car's ignition, start the engine, drive it around for an afternoon, or even a day, and then abandon it. (The rides usually stopped when the car ran out of gas. Joey was *not* going to buy gas to put in someone else's car.)

In those early days of joy-riding, Joey often took a car in order to visit a redheaded girl named Sandra, who was sixteen or seventeen years old. He had met Sandra at Fernald, just before her parents had helped her to be released. Joey had kept her address and made sure to visit her when he got out. They spent many nights in his car, parked at Revere Beach. But it wasn't long before he got involved with other young women, and that was what ended his relationship with Sandra.

Though losing Sandra hurt, Joey was so busy that he didn't have time to mope around. He worked hard at the bakery and played hard in other people's cars. He went to the beach, and to nightclubs all over the city. For a baker's apprentice, he enjoyed a remarkably flashy lifestyle.

Joey finally took one too many risks when he stole a sky-blue Thunderbird, filled it with gas, and drove north on Route 3 to visit his family in Billerica. When he arrived, his father wouldn't believe Joey's claim that he had borrowed the car from a friend. Without warning his son, he called the police, who came and phoned in a report on the car. When they heard back that it was stolen, they arrested Joey. He was convicted of auto theft and served six months in the Middlesex County House of Corrections.[11]

To Joey, the jail didn't seem much different from Fernald. It was actually less crowded and offered more privacy than any state school ward. The food was better, too. When he got out, he returned to stealing cars, eventually conspiring with attendants at city parking lots to drive off with customers' cars. (A few dollars thrown in the right direction gave him the pick of the vehicles.) Joey also learned which body shops and garages would pay him for the cars in order to scavenge their parts. The income from these thefts supplemented what Joey earned as a baker. This money became more important to him after he met Lois Ober.

Joey met Lois when he and his friends had started cruising the suburbs looking for fun at bowling alleys, soda shops, and pizza parlors. He came upon her sitting with some of her friends in parked cars in the lot of a bowling alley in Waltham. Joey bought some beer, and the young woman accepted an invitation to his apartment. Joey and Lois were drawn to each other and talked the night away.

After that first night, Joey was in love. He spent most of his spare time, and all of his spare cash, with Lois. When she revealed she was pregnant, Joey didn't care that he wasn't the father. He immediately asked Lois to marry him. She said yes. The ceremony was performed before a justice of the peace, with Fred as a witness. The couple moved together to an apartment in East Cambridge. When their daughter, whom they named Leah, was born, they got a bigger apartment in Waltham, just a few miles from the Fernald School.

Even though he stole as many as a hundred cars, Joey never thought of himself as a dedicated criminal. But he also understood that he would never make much money working as a baker in someone else's shop. Looking for a skill he could quickly turn into a business, he enrolled at a barber school. He wound up learning more from the student who sat beside him than from the instructors who taught his class.

"This one guy named Phil always had money, lots of it," recalled Joey. "One day I asked him where he got it, and he just said, 'It's simple. I'm a thief.'"

Phil told Joey that two television sets, fenced through businessmen he knew in Somerville, would fetch him $300. If he picked up a few more items—stereos, jewelry, coin collections, etc.—he could double that figure. Soon Joey was joining him on day trips to wealthy suburbs such as Sudbury and Weston, where they looked for houses where newspapers were piled up on the driveways. They would read the homeowner's name on a mailbox, then go to a corner store and ring their phone. If no one answered, they would return, break a rear window, and let themselves in.

"We would pile the TVs and stereos by the door and then go look for little stuff," added Joey. "Because it was daytime, and we had plenty of light, we could work fast and be out in a few minutes. It was pretty easy, actually."

Eventually, Joey and Phil became well known among fences and others in their line of work. Police mounted an investigation and leaned on some informants. The informants gave the police their names.

Following the lead offered by the informants, police officers showed up at Joey's apartment. He wasn't there, but Lois let them in and allowed them to search for stolen goods. (She didn't know about Joey's career as a thief and assumed she had nothing to hide.) After going through every room and every closet, the officers found just one item on their list of valuables taken in recent robberies: a driftwood statue of the Madonna. They took the statue and left Lois the name of a detective, his phone number, and a request for Joey to call him.

When he came home and heard what had happened, Joey immediately dialed the number the police had left. The detective asked him to come to the police station. Knowing that the driftwood Madonna was their only piece of evidence, Joey believed he would never be prosecuted. But to avoid a scene at his home, he went to the station.

It turned out that three different police departments had been trying for months to stop the break-ins that were occurring in the wealthy suburbs west of Boston. When Joey appeared at the stationhouse, a detective took him to a big room where hundreds of stolen items, including dozens of televisions, were displayed on folding tables. Sheets of paper, with the name Joseph Almeida written on them, were taped to several of the sets. Joey was told that prosecutors could prove he had taken them and that things would go better for him if he confessed. He refused, and then called the police a bunch of crumbs and bums.

When the trial date came, Joey entered the courthouse and immediately bumped into one of the officers who had worked on the case.

Joey would recall that the officer looked at him with both anger and bemusement and said, "You must have a lucky horseshoe up your ass, kid." He then confided that the search of Joey's apartment that had yielded the driftwood Madonna had been illegal. The warrant the officers had presented to Joey's wife, Lois, was unsigned.

In Joey's version of the tale, after the court was called to order, the judge reprimanded the police officers and announced that the charges were dropped. After the hearing ended, Joey turned to leave. But the judge stopped on his way to his chambers and called to him, asking if he wanted the statue. Joey would say, "No thanks, Your Honor, she's caused me enough trouble already."

Though he escaped jail, the arrest for the break-in convinced Joey to stop robbing houses. But he still had trouble controlling his impulses and keeping to a straight path. Where others might try to avoid trouble, Joey seemed to dive right in. A telling example of this tendency arose on a rainy night in the Back Bay, when he volunteered to fetch a friend who was finishing his shift at his job. The man's wife told Joey that her husband's car, a little Volkswagen, was unregistered and uninsured, but he took it anyway. He got to the man's shop without incident, but on the way home he made a left at a spot where a big sign announced NO LEFT TURN.

Up ahead, a police officer was standing next to his patrol car, flagging down cars that made the illegal left turn. He waggled his flashlight at Joey, indicating that he should pull over. As he pulled the car over to talk to the officer, all Joey could think about were the expired registration and insurance. The fines for all of the violations—a wrong turn, no insurance, no registration—would be expensive.

The officer who stopped Joey asked for his license and registration but then told him to move his car a few feet to a safer spot. Joey started the engine, put the stick shift in first gear, and floored the gas pedal. Since the car was a tiny VW, it didn't exactly rocket away, but the street was wet, and the car moved fast enough that Joey had to struggle at first to control the steering wheel.

The chase that ensued became the most dangerous few minutes in Joey's young, wild life. At the first red light he approached, where cars blocked both lanes, Joey leaned on the horn and just steered the Bug between them. Sparks flew as metal hit metal on both sides. The same thing happened at the next light, and then Joey turned onto a one-way street, heading in the wrong direction.

With his friend in the passenger seat screaming for him to stop, Joey finally recognized that he could kill himself, and others, and pulled the car to the curb. In what seemed like just a few seconds, police officers flung his door open, pulled him out, punched him a few times, and then pressed him hard up against its hood.

This time, Joey's luck ran out. His court-appointed lawyer was unable to devise a defense that made any sense, and there were no technical problems in how the Boston police conducted the traffic stop, the chase, or his arrest.

In a brief trial, Joey was found guilty. The judge set his sentencing and incarceration for a future court date, and Joey was released on bail. He then disappeared. Notices sent out by the court were returned as undeliverable. At a hearing attended only by the prosecutor, Joey was declared a fugitive from justice. A warrant was issued for his immediate arrest.

While in hiding, Joey called his attorney. The lawyer told him that he should turn himself in. As an officer of the court, he was supposed to tell him that. But he also said that the authorities were not likely to come looking for him. The jails were so crowded and the justice system was so overwhelmed with cases that the police tended to just wait for fugitives to mess up again and be arrested for a new offense.

Six months in county jail for stealing a car was one thing. Two years in state prison was quite another. Joey decided to call the state's bluff. He would become a fugitive and live for years in fear of a traffic violation, or a chance encounter with a police officer, that would send him away for even longer.

If Joey had wanted lessons in staying out of trouble, he might have sought out Robert and Albert Gagne. Once they were fully discharged, the Gagnes left Emerson Hospital and settled in Lowell, where jobs were plentiful. Robert worked in plants that made shoes, blankets, and even boxed spaghetti. Another former State Boy, Arthur Donovan, lived in Lowell. He introduced Robert to his sister Mary. Mary had been in foster homes, so she understood something about the loss Robert experienced as a child. They fell in love and were married in 1961. Together they would have five children, and they would still be married more than forty years later.

Discharged before his brother, Albert Gagne fell in love with cars and bought himself a string of used convertibles, trading up each year for a newer model. Like many former State Boys, he was sometimes exploited by salesmen who found it easy to talk him into high payments at high interest rates. More than one of his cars was repossessed, but this didn't dim his enthusiasm for chrome, whitewalls, and ragtops.

In his first few years of freedom, Albert had a couple of steady girlfriends, but neither of these relationships worked out. In 1961, he left Emerson Hospital and found a job in the kitchen at Wentworth-by-the-Sea, a Victorian-era resort hotel in coastal New Hampshire. When the hotel closed for the winter, he accompanied Tommy Bouchard, a friend he met at the hotel, to his home in Lewiston, Maine.

The Bouchard family sheltered Albert for a year. In that time, he and Tommy both signed up for the Army. Tommy was accepted, but like most State Boys, Albert was rejected based on a government policy that barred those with IQ scores below a certain point. (If he had waited to apply after 1965, Albert might have been accepted, trained for combat, and sent to Vietnam under a program called Project 100,000. Under this initiative, the Army accepted men with test scores as low as 62, and used their numbers to reinforce divisions in Southeast Asia.)[12]

While Tommy departed for basic training, Albert worked for a town highway crew and then in a textile factory. He finally settled into a production-line job at a shoe company called Sabelman Plastic Heel. There he met a young woman named Doris who had been raised on a hilltop farm more than twenty miles into the Maine woods. He took her to restaurants and drive-in movies. He told her about his childhood in foster care and at Fernald. Doris asked only if he was a drinker—the answer was no—and if he had ever had trouble with the law—no again.

Years after they met, both Doris and Albert would recall that they were very attracted to each other from the start. In Doris's eyes, Albert was worldly and reliable. She also thought he was pretty good-looking.

Albert saw Doris as both beautiful and kind. She made him feel good about himself.

Determined not to lose Doris, Albert went to a pawnshop in Lewiston and bought a ring. Then, during a dinner at her parents' farm, he blurted out his proposal. Doris said yes. A justice of the peace named Sadie Garland married them. He was twenty-six. She was twenty-two. They would settle in a small home on a corner of the family farm and live out their lives, far from Fernald, the Commonwealth of Massachusetts, and all the other State Boys. They had one child, a daughter named Karen.[13]

Unlike many State Boys, Albert didn't run from his past. He was candid with those who asked about his childhood, and he also tried to find his long-lost siblings and resume relationships with them. In the mid-1960s, he traced his sister Doris to a suburb of Buffalo. When they met, she was finally able to explain her disappearance in 1952.

It turned out that on that hot July day, when Doris was dropped at the foot of the Peace Bridge in downtown Buffalo, she had spent a few hours wandering downtown, trying to figure out her next move. Penniless and desperate, she finally wandered into a Catholic Church, where the priest found her a place at a home for girls called the House of the Good Shepherd.

When they heard her story, the nuns who ran the girls' home had Doris examined at a local psychiatric hospital. Doctors there found her quite sane, and capable. They contacted authorities in Massachusetts, who said they wanted Doris returned. But Doris was of legal age in New York. According to state law, she could choose to stay in Buffalo as an independent adult, and she did. In a matter of weeks, the nuns at the Good Shepherd home arranged for her to be a live-in caretaker for an elderly blind woman. She met the woman's thirty-five-year-old son and, before she turned eighteen, married him.

Although her first marriage was soon annulled, it lasted long enough for Doris to be traumatized by her wedding night (she knew absolutely nothing about sex) and to discover that her betrothed was sterile. His inability to have children would be the grounds for the annulment. Her second marriage was to an autoworker named James Testa. They would create a family that would one day include six children.

When they were reunited, Doris told her brothers all the details of her escape from Massachusetts and her life since. She also explained that she never came back for them because she was desperately afraid. Although New York officials had told her she was free, she couldn't believe that she was safe from arrest in Massachusetts. Time passed and the demands of her marriage and her children consumed her.

Doris didn't tell Albert that during stressful periods in her life she sometimes caught herself rocking back and forth, trying to calm herself as she had as a child. She didn't mention that she felt numb all the time, unable to feel the emotions that she knew others experienced. And she said nothing of her fear that something very basic, and precious, was missing in her relationship with her own children. Though she fed them, kept them clean, and treated them kindly, she didn't feel the urge to hug them and kiss them and hold them close. She was all business, like an attendant in an institution, and she knew that both she and her little ones were missing something vital.[14]

In the early 1960s, as scores of young men and women left Fernald to make their own way, some were left behind. Louis Frankowski, who was befriended by Bea Katz, was so impaired by cerebral palsy that doctors couldn't imagine he would ever function on his own. He would stay at the institution for many more years. Bobby Williams, who was separated from his twin brother Richard, was so angry and uncooperative that he was almost always under some form of punishment, which made him ineligible for the parole program.

Without his brother and his friends, many of whom were paroled, Bobby became more and more unhappy. He ran away several times and, in the summer of 1960, spent three weeks as a laborer with a traveling carnival before he was caught by state police in Southboro, about twenty miles west of Fernald.

Although he resented being kept at Fernald, Bobby considered himself more fortunate than his twin brother Richard. A leader of the 1957 riot, Richard remained in Bridgewater until June of 1961, when he passed a state-administered psychiatric exam and was paroled. He immediately found a room at a YMCA and a job as a housekeeper at Deaconess Hospital. He then hired an attorney named Charles Burgess to petition Malcolm Farrell for the right to visit his twin on Company Sundays:

> Shortly after he went to Bridgewater we were able to convince him [Richard] his only opportunity for entering society would be through a change in attitude.
>
> He subsequently learned to walk away from arguments, and he has made great strides toward readjustment but is very lonely due to a lack of friends. His one goal is to have Robert reach the stage where he can be discharged from Fernald and live with him.

Malcolm Farrell was persuaded by the lawyer's letter and agreed to the request. The Williams brothers were reunited in October 1962, when they spent four hours together in a day room on the Fernald campus. At about this time, a psychologist at the state school tested Bobby Williams and found his IQ was 76, a full twenty-five points higher than his previous score. A Fernald evaluator named Sidney Ribak found him quite capable "considering the long institutional background." Still, Bobby wasn't deemed ready for the release program. He would live at Fernald three more years before he was paroled to a job at a Brigham's Ice Cream Shop in Harvard Square and a single room in Watertown.[15]

Charles Hatch, the last of the rioters to leave Bridgewater, was finally set free on September 26, 1962. Completely unaccustomed to freedom, he would depend on his brother to help him find a job with a construction company and a room that cost $8 per week. More than forty years would pass before Hatch would again meet any of the State Boys.

By 1965, just one of the brighter men who lived with Fred Boyce in the Boys Home remained institutionalized. Thoroughly adjusted to life in the Fernald State School, Charlie Dolphus had declined to leave. (Much like the famous Mayo Buckner, he felt comfortable in confinement and dreaded the idea of being on his own.) Even when the BH was torn down, Dolphus took temporary housing until a new dorm called Kelley Hall was built and opened in its place.[16]

* * *

The parole and discharge of Fernald's State Boys, and of countless others like them at institutions across the country, would turn out to be the beginning of a much bigger change in institutions. Although policies that favored custodial care had pushed the population of America's state schools to about 230,000, opposition to the institutional approach was building. Pearl Buck's rosy images from Vineland were being replaced by disturbing glimpses of back wards. At first, the truth was revealed to individual parents, who managed to get past the day rooms to see, smell, and hear the conditions. Then came more public revelations of the life inside typical state schools. The most important truths would be leaked from, of all places, the North and West Buildings at Fernald.

By the fall of 1965, Lawrence Gomes had become principal of the school at Fernald, and he had received advanced training in special education at Boston University. The chairman of the special education department at BU was Burton Blatt, an outspoken critic of institutions who saw in them reminders of the Nazi Holocaust. For his time, Blatt was radically optimistic about the abilities of the disabled. He also believed that if informed, the American taxpayer would not allow a place like Fernald's North Building to exist.

Blatt inspired a whole generation of professionals, including Gomes, who was appalled by the overcrowding, filth, and neglect he saw when he visited many of Fernald's wards. When Blatt asked him to help sneak a photographer named Fred Kaplan into the North and West Buildings during Christmas week, Gomes was happy to help. During an extended tour of the places that outsiders never saw, Kaplan snapped pictures with a camera that he kept hidden in his clothing.

Fred Kaplan's black-and-white pictures were grainy and, since they were taken at waist height, many were only partial images. But among the scores of photos he made were more than a dozen that showed perfectly the stark reality of the institution. He captured the bent and twisted bodies on the floors, the naked men shivering in the cold, and a child, restraining straps hanging from his wrists, curled into the fetal position on a settee.

What Kaplan did not convey with pictures, Blatt covered in a text he wrote about their visits to Fernald and four other institutions in the Northeast. (At each of these schools, Blatt had the help of staff members who were ashamed of conditions in certain wards but lacked the money to improve them.) Beginning with the locked, heavy metal entrance doors, Blatt described barren day rooms jammed with adults, isolation cells where children lay in their own urine, and wards where feces were splattered on the ceiling, walls, and floor. He wrote of children whose hands and legs were tied, infants abandoned in understaffed nurseries, and attendants standing helplessly in rooms filled with as many as thirty crying toddlers.

The things that impressed Blatt were the same things that stayed with the State Boys long after they were released: the huge wooden settees, the barren tile walls, the stench. He called his essay *Christmas in Purgatory* and published it, in book form with Kaplan's pictures, in 1966. Along with his searing descriptions, Blatt reported mundane and important facts. Per-person spending at state schools averaged less than $5 a day. Six states spent less than $2.50.[17]

(Blatt could have found an exception to the stingy neglect suffered by so many residents of institutions if he had visited an elaborately equipped experimental laboratory that had been established at Fernald by devotees of the behaviorist B. F. Skinner. There a small number of children worked with a large staff of teachers, graduate-student interns, nurses, and social workers. But even where money was spent freely, the treatment of the children was disturbing to witness. The behavior-modification program involved rewarding children for good behavior—mostly this involved giving them candy—and punishing bad behavior. Punishments included tying children to chairs and confining them to closetlike spaces. One child was subjected to a week-long program that included electric shocks. The program was eventually abandoned, but the details of what occurred were recorded by a staff member, Florence Little, who was so appalled by what she saw in this supposedly advanced program that she complained formally to state and federal officials.)[18]

Though Blatt's book disturbed his profession when it was published, the general public didn't get inside the nation's snake pits until the pictures and parts of the essay were republished in 1967 by *Life* magazine. Five years later, the TV journalist Geraldo Rivera brought similar pictures into millions of homes with a documentary filmed at Willowbrook, a state school in Staten Island, New York, that housed 5,300 patients.

The exposés done by Blatt and Kaplan, and later by Rivera, helped build public support for state and federal laws that began to change the lives of people in institutions. Billions of dollars were made available in new federal programs to create community housing for institutionalized people. In many states, parents sued and won court orders that forced authorities to move people out of state schools.

The process was not uniform from state to state, and some institutions continued to hold people who were not retarded well into the 1970s. The *Miami Herald* disclosed that hundreds of people who were physically disabled, poor, or "unwanted" but not mentally impaired continued to live in Florida's state schools in 1974.[19]

Fortunately, Florida was an exception. In Massachusetts, Fernald State School was steadily emptied until only 300 residents remained. Louis Frankowski was among those who received extensive training and was moved to a specially equipped apartment, which he shared with one other person. He found employment in a so-called sheltered workshop and received help from a social worker who visited him often. Once outside Fernald, Louis learned that his parents had died. He was, however, reunited with a sister as well as with aunts, uncles, and cousins. Instead of the blank piece of paper he once kept in his pocket, he had real cards and letters from real family.[20]

Louis Frankowski's new life reflected a sweeping national trend. By the year 2002, the population of the nation's institutions would total just 47,000, compared with 230,000 in 1967. This decrease occurred despite an increase of 100 million in the overall U.S. population. Those who continued to need institutional services were the most severely disabled. Vast improvements were made in their care as per-person spending rose, on average, to more than $320 per day. Staff at many institutions, including Fernald where lone attendants once managed as many as forty children, outnumbered patients by more than two to one.[21]

The numbers do not begin to illustrate the scope and quality of the change they represent for children. In the years since the State Boys were released from Fernald, entire new disciplines have been established to identify learning disabilities and provide education in public schools. Universities began programs in disability studies. Subspecialties that focus on the disabled are now found in law, medicine, psychology, even architecture.

Viewed from a historical perspective, the shift in the status of disabled or disadvantaged children may have been more dramatic than the changes in race relations that occurred in the same period of time. It is reasonable to say that a Freddie Boyce or Joey Almeida born in 2004 would be diagnosed with a common learning disability. And he could find adequate educational assistance to succeed in school all the way through college.

* * *

As young adults, Fred Boyce and Joey Almeida did not follow closely the changes at Fernald and places like it. They were unaware of improvements made in evaluating and then helping children who had trouble learning. And they knew very little of the rapid development of special education.

Though they pushed Fernald out of their daily thoughts, the resentments they harbored over their confinement and abuse never died. Joey nursed fantasies of revenge. Once, he went so far as to discover where James McGinn lived in his retirement. He and Freddie went to the neighborhood and arranged to bump into him on the street. But they were tongue-tied during this brief encounter— McGinn had trouble recalling them—and never followed through on their daydreams about making him pay for what he had done.

As time passed, Fred and Joey were able to push Fernald further and further from their conscious thoughts. Neither told new friends where they had grown up. Joey talked about family in Billerica. Fred said he came from Merrimac but that his family no longer lived there. This lie became much easier to maintain in 1965, when he left the Back Bay and the other State Boys who lived there for a life on the road.[22]

TEN

Fred Boyce left the Back Bay in the front passenger seat of a shiny black 1963 Cadillac, the kind that had a toothy chrome grille, high tailfins, and puffy whitewall tires. In the back seat sat Stewart Aucoin and his girlfriend Margie. Up front, in the driver's seat next to Fred, was a man of thirty-three who had blond hair, blue eyes, and a salesman's way of talking. His name was Bobby Catalano.

Catalano, who spent winters driving a cab in Boston, had met Fred in the White Tower. He would rush in for a cup of coffee and, more often than not, Fred had it ready before he even asked for it. Fred had told him that he was getting tired of counter work. For a while, he had even considered joining the Army, but a friend who had been reading about the war in Vietnam had persuaded him to stay a civilian. Catalano told Fred story after story about making lots of easy money with simple games of chance he ran at carnivals and country fairs. Eventually, he talked Fred into joining him on the circuit as soon as the snow melted. A spring morning in 1964 found them both bound for Cornwall, Ontario, just north of the New York border.

They were headed for what carnies call a "still date," a small fair where the turnout is light and you only make money if you keep your expenses to a minimum. Catalano's expenses were low. His two concessions were "stick joints," small stands made of plywood, two-by-fours, and painted canvas. He carried them in a trailer that was hooked to the back of the car. At every stop, he would unload the pieces, put them together, and install the games. Customers would buy six rings for a dime and try to toss them over a peg, or use a cork gun to shoot a paper cup off a shelf.

Fred was going to be a barker for one of the games. Though he had never done this kind of work before, Catalano had been impressed by the way Fred dressed, by how easily he was able to chat with anyone, and by the way he handled customers at the White Tower. Fred was comfortable with people, even the drunks who came in when Crusher Casey's bar closed for the night. He could charm them, amuse them, and when it was necessary, he could soothe their hot tempers. These were the skills that enabled a good barker to bring in hundreds of dollars a day on the midway.

As the Cadillac raced down the Massachusetts Turnpike and into New York State, Catalano talked about living high on the road. But when the group finally got to Canada, Fred's first impressions of the carnie life were full of rain, cold, and hunger. The Catalano crew had left Boston with very little money, and the first few days of the Cornwall fair were ruined by bad weather. They were down to a few dollars when the fair manager ran into them in the dining room of the motel where they were staying.

The manager was confident that business would pick up, and he told Catalano's crew that they could eat well if they just charged their meals to their rooms. To test whether the motel staff would post the charges, he told them to give him their remaining money, and then ordered himself a big breakfast. If he were allowed to charge the meal to his room, they would get their money back. If not, they were going to buy his meal. When the bill came, he signed it, wrote his room number on it, and the waitress accepted it.

"The guy then said, 'If they do it for me, they've got to do it for you,'" recalled Fred.

For the next few days the band from Boston ran up their charges. But when the weather cleared they made a big profit on the midway because Fred, it turned out, was a very good barker. One man alone spent $30 to win a $1 stuffed toy. Fred had never felt so powerful in his life.[1]

In his first months on the road, Fred saw Indianapolis and Milwaukee and scores of little towns in the East, Midwest, and South. Actually, he saw little of the cities, since most of his time was spent at the fairgrounds and in motels. Working eight to twelve hours per day, Fred refined his approach to snaring customers. He learned how to tweak young men, charm their dates, and make himself the butt of the joke whenever necessary.

Fred also began to understand the economics of the business. In the 1960s, before theme parks and high-tech roller coasters, carnivals offered some of the biggest attractions many people ever saw. A huge Ferris wheel or flashy Tilt-A-Whirl ride would increase traffic, and therefore income, for everyone doing business at a particular fair. The same was true when big-time entertainers appeared. In their heyday, the likes of Liza Minnelli or Chet Atkins could draw tens of thousands of people.

Crowds only gave the barkers a chance to make money. Once the people were through the gate, it was up to Fred to win them over. Catalano's joints were set up on midways that might have as many as a hundred other concessions offering everything from candy apples to basketball games. People tended to stroll up and down like they were window-shopping. The free-spending types might stop at any joint and drop a few coins. But people who were careful with their money—and that meant most people—wanted to feel comfortable, and confident that they would have a fair chance of winning.

Fred put people at ease by wearing his jacket and tie, even in hot weather, and always speaking politely. (Sometimes, Catalano complained that Fred looked too good, and sounded too smooth for some small towns.) All those years spent reading the faces of attendants and other State Boys had given Fred an uncanny ability to assess people in an instant. He was friendly, and he didn't try too hard. He made players enjoy the games, even when they didn't get a prize.

In the hours when Catalano relieved him, and Fred had free time, he wandered the fairgrounds and soaked in the atmosphere. The heavy scents of cotton candy and fried dough hung over every carnival. In dry weather, dust coated everyone and everything. When the weather was wet, the mud could get so sticky it would pull the shoe right off your foot. At night, the glow of the lights reached into the sky, and the sounds of diesel generators and mechanical rides would carry for miles.

The happy sounds and sights of the carnivals made them seem like casual events, but they actually ran on a strict schedule. One of the first things Bob Catalano asked whenever he got to a new site was, "When do we spring and when do we go slough?" Most fairs would spring, or open, around noon on a Friday. Closing depended on the flow of business. If the crowd was strong on a Saturday night, the fair might not go slough until sometime after 1 A.M. Only then were the lights turned off and the gates closed.

At the very large fairs, the brightest lights came from huge concessions made out of sheet metal and covered with hundreds of colored bulbs. Some of these joints were so big they traveled on wheels and were pulled by powerful trucks. Once set up, they could serve dozens of customers at a time, pulling in many thousands of dollars per week. The operators of these big units paid high rent—fair managers charged by the foot—and their electric bills were enormous. But in the right location, they all but minted money.

Location—carnies called it "loke"—was everything. A good loke was where large numbers of people might pass by and competing concessions were out of sight. For example, if you got a spot on the right-hand side of the fair entrance, you might make twice as much money as the fellow across the way. That's because most people look to the right as they walk. By the time they return to the midway, on their way out of the fair, they are too broke to spend any money on the other side of the path.

A good loke also depended on the concessions that might be nearby. A ring-toss game could do well if it was squeezed between an ice-cream stand and a souvenir-seller. But if it was stuck right next to another basic game, few customers spent money at both. People also tended to gravitate to the fanciest attractions. Put a water pistol game next to a stick joint, and the stick joint will die. And God help you if the booth next door is occupied by a local political candidate or a Bible-pounding preacher. Both were sure to drive people away.

The best locations were hard to come by at big dates like the state fairs in New York, Iowa, and Minnesota. For this reason, when Catalano took his little games to small towns where the competition was less intense, and rents were low, he sometimes cleared more profit. For Fred, these out-of-the-way carnivals also offered a view of acts and attractions that were so old-fashioned they no longer appeared at the big towns.

One old-time joint that Fred would long remember was a sideshow billed as the New Zealand Dancing Ducks. "They dance to the music, or you get your money back!" was the barker's claim. Once a crowd was brought inside the tent, a curtain was drawn on a small stage that was really an electric griddle. A couple of strange-looking birds, probably exotic chickens, were placed on the griddle and the heat was turned up. As music played, and their feet got hot, the birds danced.

The "ducks" were just one of countless gaffs that could make a rube laugh while he lost a few pennies. At a country fair in North Carolina, Fred watched as a man set up a fake "discount" entrance that promised 5-cent admission to a fair that cost 50 cents at other gates. People paid their nickels only to discover that the real gate was on the other side, flanked by a donkey wearing a sign that read YOU'VE JUST BEEN HAD! HEE HAW! The few who asked for their money back got a cheerful refund, but most customers thought that the laugh was worth the 5 cents.

The hosts of long-running fairs in small towns presented some of the most unusual attractions Fred saw. In an isolated New Hampshire village, for example, the locals always built a tall wooden tower and hoisted an old car to the top of it. On the last night of the fair, they lit the tower on fire and people gathered to watch the thing collapse. The spectacle guaranteed a huge turnout, and the crowd would spend enough money to assure a profit for the concessions.

Little fairs were also havens for sword-swallowers, contortionists, and the kinds of sideshow acts that were becoming unacceptable to crowds in bigger cities. At one of these shows, Fred saw a dwarf whose legs were amputated draw crowds as the "girl so small she'll fit in your hand." At the same fair, he found Siamese twins posing and selling autographed photos for hundreds of people.

The human oddities often appeared beside girlie shows, where the more the audience applauded, the more a lady removed. By the early 1960s, these shows were dying out, victims of *Playboy* magazine, X-rated movies, and changing sensitivities. It didn't help that the performers sometimes caused more trouble than they were worth to a fair's sponsors. One woman actually shot and wounded her boss a few weeks after Fred saw her at a carnival in Vermont.

A shooting would give the carnies something to talk about for weeks. But such dramas were rare. For the most part, carnies concentrated on the weather and money. They talked about how much they made last week, how much they were making this week, and where they might make some next week. Every season built toward October, when the harvest was over, and fairs drew the largest crowds of the year. The last big fair of the year was always in Phoenix. In 1964, Catalano had his usually good week there, and then announced he wanted to play one more date. He had heard about a big livestock exhibition outside Los Angeles. It wasn't exactly a carnival, but he thought he might make money there.

The Catalano group drove all night to reach Los Angeles. They found the fairgrounds, set up the joint, and then spent a day in a hotel. When they came back to spring, they discovered they had established themselves in a nearly empty lot. The action, it turned out, was half a mile away beside a big arena. All the good spots were taken. It was an awful loke.[2]

In the public mind, carnivals are full of con men and criminals. Fred found the stereotype was true in a few cases. But these characters didn't last long. They would run into trouble with the police, or flimflam another carnie and disappear. The same was true for the laborers who drank heavily or took drugs. They'd fail to show up, get fired, and be replaced in a matter of hours.

The hiring and firing were done by the permanent community of traveling carnies, who always looked ahead to playing the same dates the next year and therefore wanted to avoid trouble. Devoted to their trade, these people had long-standing relationships and were slow to admit newcomers. Fred made a few friends among these regulars in 1964, and many more when he went back on the road with Catalano in 1965, proving he was serious about the business. He enjoyed the colorful chatter of the carnival community and even the nicknames—Slow Walker, Snivelin' Bobby, Steve the Puke, Blood Test Dave—each of which came with a long story.

For the most part, the carnies were shrewd, if not highly educated. Many could not read, and struggled to manage their money. Over time, Fred would discover that quite a few came from troubled families or had also been wards of the state. And like the crowd Fred knew in the Back Bay, the carnies were open-minded and didn't judge him negatively if he mispronounced a word or two. He knew he was accepted when someone started calling him Boston Freddie and it stuck. A nickname meant he was "with it," the carnie term meaning he was part of the show.

If Bobby Catalano had upgraded his games to keep the attraction fresh, Fred might have stayed with him for decades. He would have been content with a 50-percent share of what he brought in and let Catalano handle the bookings, travel, and relationships with carnival operators. But by the early 1970s, Catalano's joints were looking worn and Fred was finding it more difficult to pull in customers. New competition from fancy games that cost $100,000 or more to build began to eat into Catalano's profits.

When he came off the road in 1973, Fred was exhausted and a bit worried about the future. He didn't have enough cash to make it through the winter and had to take a few part-time jobs. To make matters even more unsettled, Catalano announced that he was not willing to invest the time and money it would take to succeed on the midway. He was through with carnivals.

The idea of taking a job where he would report to the same place every day and punch a time clock made Fred feel trapped. He liked traveling and depending on his own talents. He also disagreed with Catalano, who had concluded that stick joints were becoming obsolete. Fred believed that people would play the old-fashioned games, if only for the sake of nostalgia, if the barker made just the right pitch. Confident that he knew how to do it, he decided that in the coming spring, he would go out on his own.

With his small savings, Fred hired a metalworker to make him a twelve-foot-long trailer. He had it painted white, and equipped it with bright, low-energy lights. He then bought a couple of more challenging games. One required a player to throw a Frisbee—Frisbee was wildly popular at the time—through a narrow slot. The other was a break-the-balloon-with-a-dart game. To make the game more enticing, he stocked up on high-end prizes such as transistor radios and enormous stuffed bears.

In his first season out alone, Fred went cross-country with his own Cadillac—a ten-year-old model—pulling the trailer. Trusted old-timers got the best lokes first, so Fred often found it difficult to find a spot at a fair. Sometimes he got shut out completely. He tried to make up for his lack of pull with show managers with hustle and hard work. If he arrived at a fair to discover that a promised space was no longer available, he quickly packed up and found another carnival where he could still make money for the week. Many carnies wouldn't take this kind of initiative.

Since he was keeping all the profits for himself, Fred had a good year. He made enough to eat in a slightly better class of restaurant and to stay in better motels. He even began to date a little bit, taking women he met in the business to movies, or dancing. Tall, dark, and rather handsome, Fred had no trouble getting dates and came close to falling in love with a woman named Janet who worked the carnival circuit, too. But when she came to Boston for a winter visit, Fred felt uncomfortable and began wondering when she would leave.

Though he heard other people talk about falling in love and creating families, Fred just assumed that this was not going to happen for him. This wasn't something he thought about in a systematic way. He just felt like he was different from everyone else and that he would never live a conventional life. It was hard for him to even conjure an image of himself married and with children. But he could imagine finding a place of his own, where he could feel secure and create an ordered, predictable home life for the winter months.

Every year, Fred had a little more trouble finding a place to rent during the off-season. Redevelopment was turning the Back Bay into a much more popular, and pricey, neighborhood. Landlords in good buildings hardly ever allowed someone to stay month-to-month without signing a lease, and those who would charged the highest rent. Sometimes, out of desperation, Fred spent a few nights on the sofa in his mother's apartment, but this never went well. Once she even woke him in the middle of the night, screaming something about how he thought she was a terrible mother but had no right to blame her for what had happened to him as a child.

Finally, in the fall of 1977, Fred came off the road with enough cash to end his annual housing problem by getting something more permanent—his own home in the suburbs. He had very specific ideas about the ideal loke for his home. Boston and the Back Bay still held their allure, and Fred wanted to make sure he could drive into the city in a reasonable amount of time. He got a map and drew circles, at five-mile intervals, with the city at the center. He then made tours of the communities at five, ten, fifteen, and twenty miles out.

Besides proximity to Boston, Fred had a few specific requirements. He couldn't pay a lot of money; however, he wanted a freestanding house with enough land to give him real privacy. Good schools and stable property values were essential, and he wanted easy access to a major highway. Last, but not least, he hoped to find a place that looked homey.

Just inside the twenty-mile circle, in a town called Norwell, Fred found a little red cabin settled onto a large treed lot. The house was just a few minutes from Route 3, the highway connecting Boston with Cape Cod, and the price was right. The only hitch came when he applied for a mortgage. Though he was making a down payment worth more than half the value of the house, no bank would give him a mortgage. His job was too unconventional, and his credit history was almost negligible. Faced with losing the deal, Fred turned to the illegal and underground segment of the economy that he otherwise avoided. A $500 payoff to the right person got him a mortgage, and he never missed a payment.

Once the house was his, Fred spent the rest of his winter break acquiring furniture and curtains and everything else to make the place a home. He would never again pay rent, or wonder where he might land at the end of a carnival season.

With the security of his home, Fred felt like he had made one big step toward self-sufficiency and away from his past as a State Boy. He hardly ever thought about Fernald anymore and rarely talked about it. Most of his friends from the state school had drifted away, except for Joe Almeida. Joe spent time with Fred every winter, often staying overnight at his house in Norwell. Occasionally, they talked about the old days. Once or twice, they even drove to Waltham to have a look at the school. Though he still dreamed about it occasionally, Fred felt damaged by the time he had spent there, and he was not drawn to Fernald. Joe was different. Fernald gripped his heart and wouldn't let go.

Joe Almeida had stopped stealing cars and robbing houses, but he still had tales to tell every time Fred came off the road. In the early 1970s, for example, he had grown tired of being a fugitive and approached a lawyer about his problem with the court. According to Joe, the lawyer arranged for him to pay $400 to a judge who cleared his record. He was then able to get a driver's license in his own name and a better job at a company that made computer circuit boards.

Illegal drugs were a big part of the after-hours scene at the circuit-board factory. With all of his contacts in the darker corners of society, Joe was able to help his coworkers and even his bosses obtain marijuana. He indulged, too, and was eventually arrested with a large amount. He was able to avoid going to jail by admitting to being drug-dependent and entering a rehabilitation program.

Based at a church in Lexington, the program used the twelve steps of Alcoholics Anonymous to attack addiction, and also offered job counseling and remedial classes in basic English and math. Joe took a few tests and was informed that he possessed the equivalent of a tenth-grade education. Upon hearing that Joe had been at Fernald and received little formal schooling, the tester praised his intelligence and added, "Boy, did you get a screwing in life."

The program helped Joe to stop dealing drugs. He would never be arrested again. Over time, he stopped using, too, but pot was a hard habit to break, especially since it was the one sure way he knew to ease the rage, shame, depression, and confusion that plagued him in the years after Fernald. He never stopped thinking about how he had been abandoned, and he often dreamed of having an extended family. For a while he fixated on his birth mother, hoping that she might claim him.

Aching to make a connection with his mother, whom he had never met, Joe consulted the directories from the Boston area and called every phone listed in her last name. When he finally found her brother, the man refused to help and rushed off the line. It took Joe another year to get up the courage, and the nerve, to call him again. Having reflected on that first encounter, this time the man—who after all was Joe's uncle—relented and gave him his mother's address and telephone number in rural Pennsylvania. Joe first sent a letter, which was returned unopened. Then he phoned.

"I called and spoke to her daughter there," recalled Joe, many years later. "She said she was a wonderful mother, that she had remarried and had children and never mentioned me ever. The daughter said her mother wanted to forget the past, and I was part of the past. She wanted to protect her. I never did talk to her."

Rejected by his father on the day that he left Billerica, Joe now had to accept that he might never connect with his mother. For reasons he would never fully understand, he then found himself gravitating toward the only part of his past that remained—Fernald. Time and again, he drove to Waltham and just cruised the old campus. He would take Cherry Lane and gaze at the superintendent's house and the Administration Building. Then he would go down Maple Street, which led to the schoolhouse and the old Boys Dorm. Sometimes he would just park there and look up at the windows of the wards he had once occupied. On these visits, Joe felt a mixture of familiarity, anger, fear, and even wistfulness. He was also curious to know whether everything he thought he recalled from his time at Fernald was actually true. It all seemed too extreme, even to him, to believe.

After several visits to the campus, Joe finally felt the courage to go into the Administration Building to inquire about his records. To his surprise, he was told that he was entitled to them. He filed a formal request and was handed a stack of files to review. The clerk said she would copy the pages he wanted. Joe opened the first folder and read the top page, which was titled "Family History." It noted that he had been abandoned by his mother, that his father was a drunk, and that his stepmother, who had "an illegitimate child when she was young," treated Joe cruelly.

One page was as much as Joe could take. He closed the file and asked the clerk to copy it all. In the months to come, he would periodically open the files, read as much as he could stand, and then put them aside. The papers confirmed his memories of his time at Fernald and evoked new ones. Certain entries noted that he had been physically and sexually abused, and that some teachers and psychologists had recognized that he was not retarded. Reading that they had known that he didn't belong in a school for the retarded added to Joe's anger.

Whenever he talked to friends about Fernald, especially those who had never been there, they were sympathetic. Invariably, someone would say, "You ought to sue them," and Joey would agree. This idea nagged at him every time he read a newspaper article about someone winning a lawsuit. Eventually, his curiosity about the issue, about whether he might have a case against the state, drove him to begin contacting lawyers.

Beginning with Alan Dershowitz at Harvard, Joe contacted a number of attorneys who seemed prominent enough to sue the commonwealth. They responded to his story with terms such as "false imprisonment" and "mental anguish" but told him a lawsuit was impossible. One law firm assigned a clerk to research the case. He concluded that the statute of limitations had long run out on Joe's injuries: ". . . action against the parties involved has been extinguished. Thus, we are sorry to inform you that we do not wish to pursue this matter for you."

Though he was frustrated, Joe still focused on getting some kind of satisfaction. He decided that a job at Fernald, with good pay and state benefits, might do the trick. On a summer day in 1984, he drove to Waltham and filled out an application. He decided not to mention on the paper that he had once been a Fernald resident. Weeks later, he was hired.

This time, instead of working inside one of the buildings, Joe was assigned to the motor pool. He drove the trucks that made deliveries at various buildings and the buses that moved patients, many of whom were in wheelchairs, from one place to another. On some days, Joe would drive for an hour in the morning and then just wait to follow the same route in the afternoon. During these long breaks, he was free to go anywhere he liked, even up to Ward 22, which had been closed to patients for years. Alone in these places, Joe would recall his past but also reflect on his adult life. He reminded himself that he was safe, and that Fernald was paying him well to travel the paths he once walked as a State Boy.

Eventually, Joe's explorations, and his familiarity with the school, caught the interest of his coworkers. They had trouble understanding how he knew every inch of the place without any assistance. When he finally told his coworkers the truth about his past—that he had *lived* in Fernald's dorms—most had trouble believing him. He was nothing like the patients they saw in the institution, and the thought of a person of his intelligence and ability spending his childhood in Fernald appalled them. One supervisor who was especially disturbed by what Joe told him about Fernald in the 1950s tried hard to think of a way to ease his pain. One day, he simply handed Joe a passkey that would get him into most of the buildings on the campus. He told Joe that he was free to explore the campus on his own and that this might help him reach some sort of peace with his past.

In the years that followed, Joe used the passkey to unlock doors at the old laboratory and the school. One night, he even went into the attic of the Administration Building, where he switched on a light and discovered scores of boxes filled with thousands of files. He searched for records bearing his name but found none. Nevertheless, he helped himself to a few souvenirs, including an oath of allegiance signed by Mildred Brazier, the one-time principal.[3]

While Joe returned to Fernald to confront his past, Fred Boyce did whatever he could to erase his. Determined to dust off any evidence that he had missed most of the basics children learn in school, he found a private English tutor who agreed to meet with him for an hour each week during the winter. She taught him about vowels and consonants, verbs and nouns. Between lessons, he read everything he could get his hands on, from menus to newspapers to books. He kept a dictionary and never passed a new word without pausing to look it up.

As he became more comfortable reading, Fred acquired a literary taste that ran toward cosmology, psychiatry, and philosophy. He read Nicolai Gogol's *Diary of a Madman, Cosmos* by Carl Sagan, and many of Isaac Asimov's works. He read a biography of Albert Einstein, academic books on the human brain, and much of Somerset Maugham.

In Maugham's books, Fred found familiar characters—underdogs struggling against the powerful—as well as themes that appealed to his sensibilities about justice and fairness. But just as in school with Mr. Bilodeau, he was most fascinated by science. He loved to read about the vastness of the universe and theories of its origins. He was extremely curious about the principles that govern energy and matter. Slowly, he developed a personal philosophy based on a few basic ideas: that each human life is precious, that science has explained just a tiny fraction of reality, and that change is the only constant.

Where once Fred was a devout Roman Catholic, he moved away from the church and traditional, organized religion. Reflexively resistant to authority figures, he was suspicious of preachers who demanded big donations and followed rigid theologies. He continued to believe that human existence depended on an unseen, beneficent power. But he became convinced that mere human beings were incapable of understanding it.

This reverence for the unknown led Fred to a remarkably generous perspective on his past. Despite all that had been done to him in his childhood, he had trouble blaming those who had abused and neglected him. He recognized that his father had been so depressed that he committed suicide. His mother was an alcoholic and obviously incapable of caring for anyone, perhaps even herself. He thought that all his foster parents, doctors, teachers, and even the Fernald attendants had been powerless in the face of a state system that had been set up to corral unwanted boys and institutionalize them. They were victims themselves, he thought. Any evil, as far as Fred was concerned, resided in bureaucracies, in circumstances, and ways of thinking that were beyond the influence of any one person. "Almost anyone could get caught up in something terrible," he would say. "It's easy for people to go off track."

In his most philosophical moments, Fred accepted that some unseen power had set him on his life's path. He believed in free will, at least when it came to self-improvement and financial security. But he suspected that fate had already determined that he would always feel slightly outside the social mainstream. He still had trouble letting go and having fun the way other people seemed to do. And the older he got, the more certain he became that love, the kind of love that makes people get married and have children, was not in his future.

Fred was a confirmed bachelor, happy enough on the road during carnival season and in his cozy house every winter. At home, he followed a certain routine: coffee with friends at Dunkin' Donuts, household renovation projects, the occasional night out in Boston. Though he got lonely sometimes, because real romance seemed to have passed him by, he also felt quite content. He stopped looking for the right woman.[4]

Abra Glenn-Allen was the daughter of a single mother whom Bobby Catalano had known for many years. The mother, Sherrill Glenn-Allen, lived in Oklahoma, and whenever they passed through on the carnival circuit, Bobby brought Fred to her house for a visit. Catalano once stayed for a few months at Sherrill's house, and her children— Abra and her sister Raina—thought of him and Fred as close family friends. Fred, who brought them toys and would even go frog hunting with them, was a favorite.

The Glenn-Allens were a closely bonded but independent family of women. Sherrill refused to rely on men to support her and taught her daughters to think and act for themselves. Her eldest learned the lesson well. When Abra was fifteen, and her mother announced she was moving to California, she married her high school boyfriend so that she could stay in Oklahoma.

The marriage worked long enough for Abra to get her high school diploma and attend college for a year-and-a-half. At age eighteen, she was divorced, and she went to Florida to find the father she had never met. She found him, but he didn't offer her the love and support she craved. She then began to nurture a new dream of settling in a big northern city, like Boston, and going back to school. She called Catalano, who told her to come north and he would take her in. But on the day she arrived, he met her at the airport with the news that he had no place for her to stay. After a flurry of phone calls, Fred agreed that she could take the extra bedroom in his house.

To Fred, the young woman who carried her life in two suitcases into his house that day seemed lost and vulnerable. Slim, brown-eyed, and dark-haired, she was practically alone in the world, as he had been when he left Fernald for that single room on Harris Street in Waltham. He told her that his spare room was hers for as long as she wished "and no one will ever bother you there."

Abra was moved by Fred's kindness. He was loyal, steadfastly dependable, and completely guileless. He told her about his past, and though it was hard for her to imagine him being diagnosed as retarded, she could see that he had a few odd mannerisms. For example, he often got mixed up when using slang. One instance that stuck with her was the time he complained that a friend lived "way out in the dune box" instead of the "boondocks."

These little quirks only endeared Fred to Abra. She understood more about the life he had lived than most other women could. For a time, her mother had been the principal of Oklahoma's state school for the retarded in the town of Enid, and Abra had worked in the place through a couple of summers. She had met seemingly normal students—some of whom were quite witty—and she had been puzzled about their assignment to the institution. With this experience in mind, Fred's tale didn't seem outlandish.

Abra also recognized that Fred had a mournful relationship with his past. He was painfully aware that his institutional upbringing was sometimes evident in his speech and his attitudes. Worst of all, Fred couldn't reach any sense of resolution about the loss of his childhood. He felt violated and wounded, and there was no way for him to find justice. She could see in him the pain of a thousand wrongs never made right.

Living together, the young woman and the older, unattached man quickly felt an attraction. Within six weeks, Abra knew she wanted more than friendship. Fred felt the same way, but was so concerned about the trust he had developed with her that he waited for Abra to say something first. She did, and they began a love affair.

It was not as much a torrid romance as it was an alliance built on mutual kindness, respect, and comfort. Fred loved Abra's intelligence and energy. He admired her drive to get an education and to make something of herself. And he didn't mind at all that she was beautiful.

For her part, Abra had never been involved with someone so sincere and generous. Though she felt a little strange about their age difference—she was twenty-one and he was thirty-eight—he looked and seemed much younger than his age, and she quickly overcame her doubts.

In many ways, they were well matched. They loved being silly together. Abra would make her eyes bug out and stalk the house as "Crazy Susan," a ridiculous character who made Fred double over in laughter. She could also be sentimental, and once came home feeling full of love for him and announced loudly, "Freddie Boyce, I think I'm going to be with you the rest of my life."

Abra was also good at coaxing Fred to do things that he considered too sentimental or corny. Once, she packed a picnic and insisted, with him resisting, that they drive to the seashore near Gloucester and spend the day in the sun. Fred thought she was crazy, and that picnics were for *normals* and not for refugees from Fernald. But the day in Gloucester turned out to be glorious, the sandwiches tasted better than any he had ever had, and the memory would be permanent and golden.

Together they continued to renovate and decorate the little house in Norwell. They once went to an estate sale, and Abra almost felt sorry for the owner as Fred talked him into selling a dining-room set he had priced at $1,500 for just $800. She was even more amazed when Fred got the man to throw in a washing machine and clothes dryer, so long as Fred unhooked them and took them away in his truck.

But as much as they loved each other, and enjoyed nesting, Fred and Abra still had to find a way to live together day by day. Always guarded about his privacy, Fred had bought his home for the purpose of living exactly the way he wanted. He had trouble compromising, and Abra found him to be a little high-strung and a little fussy. A good example of this tendency arose when Fred bought a table for their tiny kitchen, one that could be attached to a wall. She came home after it was installed and said it was in the spot she had reserved for a stove. He wouldn't move the table until she cried with frustration.

The spat reflected a bigger problem. Abra felt that Fred was slow to acknowledge her feelings and reluctant to talk about their relationship.

He was fine when it came to practical matters and abstractions like science and politics. But when emotions were involved, he used John Wayne as a role model. He didn't ask her questions that showed he was interested in her feelings, or to help her elaborate on them. She knew this was a legacy of Fernald. Fred had never seen anyone talk in an intimate way. But understanding this didn't make her feel any less lonely.

Though he tried to talk in the way Abra wanted, Fred found it a strain. Once, when she complained that he never complimented her, Fred grew quiet and studied her face. He then said, "Abra, you are a really beautiful woman. Your eyes are so evenly matched." She couldn't help laughing. Fred felt more frustrated.

"I would say to Abbie, 'Can't you see by what I do and how I act that I love you?'" recalled Fred. Abra could see it, but she wanted more. And she believed that over time she and Fred would be able to talk in the way she wanted.

In the meantime, Abra took to the road with Fred, living in a camper-van that he had bought to pull his concession. She tried to learn his trade and even took some shifts barking at customers on the midway. But she was no Boston Freddie, and she knew she never would be. She was happier just to sit on the side, helping with the prizes and the cash, and being good company. Besides, she had her sights set on college.

Two years after she came to Norwell, Abra enrolled at the University of Massachusetts in Boston, where she would earn both an undergraduate degree in English and a master's degree in the teaching of English as a second language. In those years, Abra made friends among her fellow students and sometimes brought Fred to campus events. One produced a memory that they both would hold for years.

The event was a lecture that had something to do with America's involvement in the affairs of Nicaragua, where the Communist-leaning Sandinista government was battling rebels—called Contras—who had received U.S. help. The aid to the Contras produced a scandal in the Reagan administration and outrage among campus liberals nationwide. Fred listened patiently to a speaker denouncing the United States, and then, when the man asked for questions, Fred raised his hand.

After being recognized, Fred stood and in a clear voice asked why American aid to the Contra rebels was worse than Russian aid going to the Sandinistas.

The murmurs and stares let Fred know that this was not the kind of question the crowd appreciated. Abra pulled on his arm, so he sat down to hear the speaker's reply, which had something to do with the factions vying for power and who was in the right. Fred wanted to follow up with another question, or comment, but Abra gripped his arm and he knew he should keep quiet.

The experience at the lecture made Fred feel wary of Abra's academic world. He suspected that some of her friends disapproved of their relationship, but the trouble was worse than he knew. Abra had been having an affair with a fellow student. Though she kept the relationship secret, Fred sensed something was wrong. Still, he supported her and couldn't have been more proud when she got her bachelor's degree in 1984 and master's in 1986. When Abra said she wanted to study further at the University of Massachusetts, he put his foot down. He thought she should transfer to Harvard. She applied, was accepted, and earned her second master's in a year.

By the time she finished at Harvard, it would have been clear to anyone that all of Abra's education was her way of preparing for a life that stretched beyond Norwell and the little red house, and perhaps beyond Fred. Abra understood this most of all. Though she loved Fred, she noticed that she never thought about growing old with him. "I actually felt like I wasn't finished growing," she would say years later, "and I wasn't ready to settle down."

At the same time, Fred tried to believe that he had found the woman who would help him defy fate and actually have a normal family life, perhaps including children. But he was not entirely certain. "When we were in bed, she'd get her arms around me and hug me with her head to my back," he recalled. "I always liked it, and I would tell myself, 'Savor this, because it may not last forever.'"

Fred's doubts were growing when Abra announced that she wanted to spend a year abroad in the Peace Corps. Sensing he had to make a move, or lose her, Fred asked her to marry him. Abra denied her feelings of restlessness and worry and focused on the affection and gratitude she felt for Fred and said yes. On February 14, 1987, Valentine's Day, they were married at the Unitarian Church in Norwell.

Within four months of their marriage, Abra applied for the Peace Corps, hiding the fact that she had a husband, which would have disqualified her from serving. With her language and teaching background, she was an ideal candidate, and she was accepted, trained, and assigned to Morocco, where she would serve for a year. The goodbye was difficult, but Fred believed the time overseas would be good for Abra. Besides, he would never do anything to make her feel trapped or unhappy in their relationship. As she left, they made plans for him to visit halfway through her term.

When the carnival season was over, Fred flew to Morocco to spend two weeks with his new wife. They traveled the countryside by railroad, from Marrakech to Tangier. Besides seeing the sights and visiting with other Peace Corps volunteers, they went to an orphanage where Abra and Fred were both taken with the sight of scores of homeless children.

After Fred went back home, Abra revisited the orphanage and filled out papers to adopt a child. Maintaining the fiction she had created for the Peace Corps, she told the authorities that she was single but capable of raising a child on her own. While she waited for the bureaucracy to work, Abra also began a romance with another American Peace Corps volunteer, Peter Figueroa. That she was able to fall for another man made her accept that her marriage to Fred was ending. When she became pregnant, she knew she would have to confront the issue the next time she saw him.

It happened that the Moroccan orphanage granted Abra's application and gave her a baby boy she called Julian. With a new son, and a baby on the way, she was feeling confused and torn. Her future with Pete, who didn't know about Fred, was in doubt. And she couldn't imagine how Fred would react when he saw her with a baby in her arms and another one clearly on its way. In July 1988, she carried Julian on the exhausting trip from Morocco to Boston and then got on a bus to Middletown, New York, where he was playing a fair.

The moment when she appeared at Fred's concession in Middletown, obviously pregnant and carrying both Julian and her belongings, was one of the most difficult in Abra's life. She watched the pain cross Fred's face, followed by confusion and hints of anger. He had selflessly supported her education and her year abroad. He had been looking forward to her return for months.

To have Abra come back to him carrying another man's child was too much for Fred. First he said he couldn't understand what had happened, and then, lost for words, he wept. He still loved Abra and had built his life around the fantasy of growing old with her. She still loved him, but not in the way a woman loves a husband.

That night, Abra, Fred, and Julian stayed in his camper in Middletown. Neither one of them would recall all of what was said. But they would forever remember how they felt. Fred was both disbelieving and angry. The life he had imagined with Abra had been lost. Lost, as well, was his hope that he might have a loving family, like other people. Abra felt confusion, sadness, and remorse. She had known for some time that the communication gap she felt with Fred, what she saw as his inability to express his feelings, would be too much to overcome. But she never expected their relationship to end this way.

When she left Middletown, Abra took Julian to Oklahoma, and her mother's house, where she would make a new start. She grew to see her time with Fred as a moment when she stopped drifting in life, trained her mind, and began forming her own identity. Though she might have been able to accomplish it all on her own, Fred had been an enormous help. He had had confidence in her, and expressed it, during times when she did not. In this way, and many others, he had saved her from the drifting that could have carried her all the way to middle age.

While Abra moved on, Fred held to the carnival routine. He played the big midsummer fairs, and then scrambled to make as much money as possible before October and the cold weather sent him back to Boston. At home, the dark New England winter settled upon him, and so did a severe depression. He felt so sad, and unmotivated, that he couldn't imagine taking to the road again in the spring. He went to his doctor on the chance that he might get some medication that would help.

"After he listened to me, he said that I was grieving, like I would if someone had died. He said that he would have been worried about me if I hadn't gotten depressed. He said I should wait it out a little, and I did. In March, I felt good enough to go out, but I would say that I was pretty down about what had happened for two or three years."

In those years, Fred would reach certain conclusions about himself and his place in the world. He thought of himself as someone who had been forced to study "normal" and therefore reacted mechanically to life, thinking about each situation and then consciously deciding how to feel, how to talk, how to act. At one point, he found himself writing down the word "spectator" as the most succinct definition of his life. "I accepted I have a life that's always going to have pieces missing," he one day explained. "I'm not unique with that. A lot of people feel that way. But I can't help but think that, without Fernald, I would have made a much bigger contribution."[5]

ELEVEN

On the day after Christmas 1993, Fred Boyce joined the millions of Americans who made their way to shopping centers for postholiday bargains. Sears had advertised a furniture sale, and Fred wanted a new chair. For $279, he got a high-backed model covered in green fabric. He carried it out to his van, where it easily fit in the back. He closed the rear door and then climbed into the driver's seat up front.

When Fred started the engine, the radio came on. It was tuned to a local news station, the kind that broadcasts every story with the same tone of urgency, so Fred didn't pay much attention as he began to navigate the parking lot. But when he heard the words "Walter E. Fernald State School," he took his foot off the gas pedal. The announcer then said something about a "radiation experiment" involving doctored oatmeal. Fred steered into a parking spot, turned off the motor, and just listened.

In less than a minute, the radio reported that a group of boys at Fernald had been fed hot cereal laced with radioactive calcium as part of a scientific experiment. Fernald doctors, Harvard University, the Massachusetts Institute of Technology, the Atomic Energy Commission, and Quaker Oats were all linked to the project, which was conducted in the late 1940s and early 1950s. The subjects had been called the Science Club.[1]

Fred immediately recalled being selected for the special group of "smart boys" who had been segregated in a day room at the BH and fed separately from the others. He also remembered how his blood, urine, and bowel movements had been collected, and how he and the others were rewarded with presents, and a Christmas party at MIT.

After he left Sears, Fred went to a Dunkin' Donuts shop where he got coffee and sat to read an article about the Science Club that appeared on the front page of the *Boston Sunday Globe*. In the piece, reporter Scott Allen quoted experts who said the radiation doses the boys had received were probably not harmful. But the report also made clear that none of the participants, or their parents, gave informed consent. A letter to parents in which Malcolm Farrell described the club's activities mentioned that the boys would get a special, enriched diet. But it said nothing about radiation.

Though the news reports brought Fernald back into Fred's life in a very sudden way, he wasn't especially surprised by the revelations in the *Globe*. Worse things had happened at Fernald right out in the open. And while he was disgusted by the experimenters' deceit, this was not the overriding emotion he felt. Instead, he was excited and optimistic. At long last, he thought, maybe the world would become interested in the truth about what happened to him and all his childhood friends at the hands of the commonwealth.

"I couldn't believe that this story about the radiation was the big deal," explained Fred years later. "But I knew this was my chance to let people know about the other story. I had spent my life hustling around the carnival business, trying to be a good person and all. But here was my chance to do something more, to make a contribution."

That night, Fred decided to call some local TV stations and offer their news reporters an interview. They could hear about Fernald and the Science Club from someone who had been part of it all. The first reporter, at a station called WCVB, brushed him off, but John Dougherty of WBZ listened. Dougherty was leery of Fred, wondering if he was mentally unstable, or had some unstated motivation for seeking publicity. But he always looked for ways to put a human face on a big story, and Fred was offering to serve in that role. They arranged to meet in the morning in Norwell at Fred's hangout, the Dunkin' Donuts shop.

The next day, Fred dressed in a new sweater and slacks, threw on a black wool coat, and drove to the coffee shop. He was alone, on a stool, when a van decorated with the WBZ logo pulled into the lot. Dougherty's cameraman got out and came into the shop. He looked around and decided that the well-dressed man at the counter couldn't possibly be a Fernald State School alumnus. He went back to the van and began to tell Dougherty that their interview subject was not inside. Fred put down his coffee and rushed out to the parking lot.

"You looking for Fred Boyce?"

"Yeah."

"You found him."

Having made his point—that he wasn't some broken-down man who lived in a hovel—Fred introduced himself to Dougherty, who recognized immediately "that this wasn't some unstable guy." The reporter and his cameraman followed Fred's car to his house. Once inside, Fred sat in his new green chair and answered an hour's worth of questions. Not yet ready to be publicly recognized, Fred was photographed with his face obscured by shadows. In the report that aired, he said that Fernald's doctors had considered children at the institution their "personal property" and disregarded their civil rights. The piece included Fred's comments about civil rights. But when he saw the broadcast, Fred felt that the truth of Fernald remained unsaid and that he needed to do more with the opportunity that the radiation scandal offered.[2]

One source behind *Globe* reporter Scott Allen's original article on the Fernald experiment, which had set off a frenzy of media coverage, was a sixty-one-year-old librarian named Sandra "Sunny" Marlow. A cultured, well-read woman, Marlow was a fine artist by training but needed other work to pay her bills. In 1991, she was hired by Fernald administrators to organize the school's collection of books and papers on mental retardation into a functional library. (Since Fernald was the oldest such institution in the country, its holdings promised to be valuable.) The old laboratory building, where Dr. Benda had conducted autopsies, was converted to house the collection and renamed for Samuel Gridley Howe. Marlow set to work on the thousands of items brought there from across the campus.[3]

On a spring day in 1992, Joe Almeida appeared at the library door. He had finished his morning bus route and had five hours to kill before his next assignment. He was curious about what was going on in the old Southard Lab. Sunny Marlow invited him in and explained her project. Like everyone else, she had trouble believing that Joe, a well-spoken middle-aged man with graying hair, had been a patient at the institution. But she could see that he still felt angry about his years at Fernald and that he harbored a powerful desire for revenge.

Unbeknownst to Joe and to the people who had hired her at Fernald, Sandra Marlow had a long-standing interest in radiation and government secrecy. Her father had been involved in various Cold War nuclear bomb tests when he served in the Air Force. In her adult life, Marlow had been captivated by the mystery around her father's work and—after he died of cancer—frustrated in her attempts to get an accurate accounting of his exposure to radiation. When federal officials explained that his records had been lost in a fire at an archive in St. Louis, Marlow dedicated herself to finding other sources for the information she wanted.[4]

In time, Marlow became an amateur expert on the government's Cold War activities and the practice of official secrecy. She joined the National Association of Atomic Veterans, which was full of people who shared the same interest, and spent countless hours reading obscure documents, books, and reports. She was always on the lookout for information about radiation, medical research, and Cold War science.

At Fernald, Marlow never expected to find materials linked to the Cold War. Instead, she anticipated long months of work on academic and bureaucratic records. The volume of materials in the Howe Library quickly overwhelmed her. But when Joe suggested she tour abandoned campus buildings where more files and books had been left moldering, she immediately agreed. Marlow was devoted to doing a thorough job.

Joe took Marlow to Ward 22, showed her the isolation cells and the day room, and helped her find files in an abandoned office. They visited the Boys Dormitory and the old gymnasium. Joe even took her into the attic of the Administration Building, where the dry heat was making the papers of Walter E. Fernald as brittle as November oak leaves.

In their meandering, Joe told Marlow stories she found both disturbing and fascinating. He described his work in the laboratory basement, slicing up brains amid the big jars that held organs and fetuses. He recalled weeding the gardens in the summer and working in the bakery in winter. He even told her a story about Malcolm Farrell's interest in his sex life. "He asked me how my sex life was," said Joe. "I said, 'Fine doc, how's yours?'"[5]

As she accumulated anecdotes, Marlow hunted for evidence to confirm the dark picture Joe painted of his time at Fernald. She collected documents about the institution's use of insulin coma therapy, electroshock, and lobotomy. She found letters written on stationery from the Atomic Energy Commission—a predecessor of the Department of Energy—and papers from obscure academic journals that seemed to reference medical experiments done at Fernald. Many were authored by Clemens Benda. She was surprised to learn that he had collaborated on studies using radioactive iodine, calcium, and iron. She also found a "Dear Parent" letter dated May 1953 and signed by both Malcolm Farrell and Clemens Benda:

Dear Parent,

In previous years we have done some examinations in connection with the nutritional department of the Massachusetts Institute of Technology, with the purpose of helping to improve the nutrition of our children and to help them in general more efficiently than before.

For the checking up of the children, we occasionally need to take some blood samples, which are then analyzed. The blood samples are taken after one test meal containing a certain amount of calcium. We have asked for volunteers to give sample blood once a month for three months and your son has agreed to volunteer because the boys who belong to the Science Club have many additional privileges. They get a quart of milk daily during that time, and are taken to a baseball game, to the beach, and to some outside dinners and they enjoy it greatly.

I hope that you have no objection that your son is voluntarily participating in this study. The first study will start Monday June 8th, and if you have not expressed any objections we will assume that your son will participate.

Sincerely yours.

CLEMENS E. BENDA, M.D.
Clinical director

Approved_____

MALCOLM J. FARRELL
Superintendent

By the summer of 1992, Marlow had gathered enough information to conclude that children at Fernald, and at the state school for the retarded in Wrentham, had been purposely exposed to radiation, apparently without proper consent. Parents had been told that the Science Club was important. They had not been told that its members would be fed radioactive elements.[6]

The phrase "human guinea pigs" came to Marlow's mind as she gathered some of the documents and drove to Brookline for a meeting with a friend named Daniel Burnstein. Well known among liberal-leaning activists in the Boston area, Burnstein was the son of a union organizer and had attended the University of California at Berkeley in the 1960s. He had never lost his generation's skepticism about the powerful. An activist lawyer, he was also president of a small organization called the Center for Atomic Radiation Studies. This group kept an eye on the nuclear industries and those who regulated them, and offered testimony on legislation and environmental issues.

Though viscerally suspicious of radioactivity and government science, Burnstein was realistic about the health effects of radiation. Low doses of the type the State Boys had received were not generally regarded to be harmful. Some scientists believed it may even be healthful. He noted this as Marlow showed him the experiment protocol and progress reports. But whatever the health risk, Burnstein recognized that the state and its scientists were legally bound to obtain informed consent from experiment subjects, and here Marlow held the kind of document—the vaguely worded "Dear parent" letter—that lawyers call a smoking gun.[7]

As he read the records, Burnstein was amused to know that the state had hired Sandra Marlow to work on the Fernald library. Few people were more likely to be as alarmed as Marlow was by an inside view of the institution's past, and fewer still would have been so determined to find damaging documents. The state, MIT, and Quaker Oats would come to regard her as the librarian from hell, thought Burnstein. In December 1993, he would help her establish this reputation by putting her in touch with the *Globe* reporter Scott Allen.

Beginning with the documents she had shared with Burnstein, Marlow showed Allen the evidence of the Science Club experiments. Armed with these papers, he then contacted experts on health, radiation, and ethics. Many of them pointed out the ethical problems of using state wards, the hazards of low-level radiation, and the obscure wording of the consent letter. Others, however, said that in the context of its time, there was nothing remarkable about the oatmeal experiment.

One of the MIT researchers, Constantine Maletskos, insisted that the studies were harmless. "I feel just as good about it today as the day I did it," read the quote of Maletskos that was published in the *Globe*. "The attitude of the scientists was we're going to do this in the best way possible. . . . They would get the minimum radiation they could possibly get and have the experiment work."[8]

Maletskos had been a coauthor of some of the papers published as a result of the Science Club experiments. In a later interview conducted by historians for the Department of Energy, he would explain that the Fernald boys were used because researchers had established a relationship with the school. Captive subjects were best, he added, "because in all of these experiments you have to have control of the subjects. You can't just let them walk around; you have to collect 100 percent of their excretions, you have to see that they're eating properly, and all this kind of thing."[9]

In the week that followed his first article in the *Globe*, Allen found new angles on the Science Club story. A professor at Brown University called attention to a 1950 memo sent to the AEC warning that human experiments were ethically unacceptable. (This communication was seen by some to indicate that the researchers at Fernald should have known what they were doing was wrong.) The Massachusetts Department of Mental Retardation announced it would investigate the medical research, and MIT released records showing that the Fernald and Wrentham projects were much bigger than previously reported and involved 120 children.[10]

From late December 1993 to February 1994, the press in Boston reported new developments in the scandal almost daily. The scope of their investigation broadened to reveal Cold War–era experiments in which hospital patients and prison inmates were given hallucinogenic drugs.[11]

At the same time, reporters in cities across America were turning to newly available government documents to find evidence of scores of

AEC projects in which thousands of people had received doses of radiation. In some of these experiments, the human subjects were enrolled in medical research. In others, government scientists simply released radiation into the environment to see its effects. Although a few of the people who were exposed understood the risk and granted consent, the majority did not.

As the revelations piled up, America's radioactive past became the subject of newspaper editorials, political cartoons, and TV talk shows. The publicity moved Secretary of Energy Hazel O'Leary to suggest that research subjects be compensated, and President Clinton created a federal committee to investigate the government's role in the use of radiation on human subjects and propose a response. Clinton promised to end fifty years of federal secrecy by declassifying thousands of documents.[12]

In the meantime, Senator Edward Kennedy and Congressman Edward Markey announced they would conduct hearings at Fernald. And Massachusetts officials announced that they were going to authorize an official report on the experiments at Fernald and Wrentham. A task force of experts would be created to gather and review information on radiation experiments at all state schools. They would then publish an official record, which would include findings of fact.[13]

When they heard that the commonwealth was going to produce its own version of the Science Club story, Joe Almeida and Fred Boyce got worried. Having visited the attic of the Administration Building, Joe suspected that important and incriminating documents lay waiting in dusty boxes. But neither he nor Fred trusted that the state would discover, preserve, or make any of it public.

Joe and Fred were not reassured when the state named the members of the panel to conduct its inquiry. Six of the ten either held or once occupied government posts. This connection made the former State Boys skeptical. Determined to do their own research, Fred and Joe went to see Sandra Marlow at the Howe Library. They found her working among boxes and documents that were spread out on tables and piled on the floor. They said they wanted to talk about the Science Club, and they stood for more than an hour, telling her what they could remember about the meals they had been fed and the attendants and doctors who conducted the experiment. Fred was especially detailed in his descriptions of the meals the club ate and of the rewards they received.

In those moments when Marlow seemed gripped by Fred's stories, Joe drifted around the library to look at papers and files. He grabbed a few documents, slipping them under his coat, but mainly he was interested in the layout of the place. He had it in mind to return.

In the course of their conversation, Fred and Joe learned that instead of commissioning Marlow to dig deep into the records, the state was reducing her hours. The official task force would soon move into the library and Marlow would be pushed out. She was upset about being replaced and also feared that the state would rush to conclude the review in order to end the controversy. Before he left with Fred, Joe wandered toward one of the windows, looked out at the snowy landscape, and silently unlocked the window.

Days after they visited the library, Fred and Joe returned to Fernald in the darkness of a moonless night. They sat in Fred's truck and waited until it seemed no one was around. They then crept to the unlocked library window.

While Fred kept watch, Joe pushed up the sash and then hoisted himself inside. Fred followed, joining Joe, who was crouched on the floor. He reached up to shut the window. For a moment, he thought about the irony of breaking into an institution he had once fought so hard to escape. Then Joe reminded him there was work to be done, and so, with his heart pounding, he switched on his flashlight.

Joe and Fred felt their way among the piles of musty books, files, and papers. They opened logbooks where they saw their own names in the notes about the comings and goings of staff and patients. They found the yellowed pages of a report on the riot at Ward 22, and disciplinary letters sent to attendants who had abused their charges.

That night, the two men sifted through thousands of pages, looking for anything that supported their memories of Fernald. They filled several boxes and stacked them by the window. They then slipped out with what they found. Included was a typewritten list of thirty-five names of State Boys including Fred Boyce and Joseph Almeida. At the top it read, "Permission received for participation in project." At the bottom was typed, "Walter E. Fernald School Science Club."[14]

Eighteen days after the *Boston Globe* made the Science Club known to the world, Senator Edward Kennedy walked into Howe Hall at Fernald. It was the same auditorium where the State Boys had watched Saturday movies and the Christmas pageant where Doris Gagne sang with her brother Albert on her lap. Now it was set up for a public hearing. Kennedy took his place at a table in the front, facing a crowd of several hundred. After he was joined by local congressman Edward Markey, the senator used a gavel to quiet the crowd. The patrician voice and the great mass of silver hair were familiar to everyone in the hall and everyone watching a live cable TV broadcast. He twisted a pen in his hands as he began to speak.

"We want to know what was done in Massachusetts and in every other state where these experiments were conducted," announced Kennedy. "We want to know what records exist, how great the dangers were, how much consent, if any, was obtained, how much harm was done."

Expert witnesses including doctors, ethicists, and lawyers briefed Kennedy, Markey, and the public on the health hazards posed by the Science Club experiments and the ethical issues surrounding what had happened. The witnesses agreed that the use of state wards, who were not fully informed, was ethically wrong. "The residents were used because they were convenient," said law professor George Annas of Boston University. He added that subjecting the State Boys to the experiments was wrong "even by the standards of the time."

None of the other witnesses at the hearing challenged Annas on the propriety of the experiments. The experts also seemed to agree that the health risk posed by the radiation was almost zero. "I don't think that anything hazardous occurred," said Bertrand Brill, a specialist in nuclear medicine. David Lister, dean of research at MIT, said that the subjects who consumed the radioactive oatmeal would have only a slightly raised risk of cancer.

While Fred watched the hearing on TV in his home, Joe Almeida stood in the back of the auditorium through the entire hearing. He was most interested in the two men who appeared to represent Science Club members. He didn't recognize their faces, but when their names were read, he knew he had lived with them at the BH. Charles Dyer and Austin LaRoque were, like him, the type of boys who had never belonged at Fernald in the first place.

LaRoque and Dyer were in their early fifties and, with their brown hair and halting speech patterns, could have passed for brothers. Each of the men tried to describe something of life at Fernald and how it had affected them. LaRoque explained that he was not properly educated and even in adulthood was accustomed to doing "what I was told to do when I was told to do it." The only moment of drama in their testimony came when LaRoque challenged Dr. Brill, asking, "Would you have allowed your son to participate in this study?" Brill, to the surprise of many in the hall, answered, "Knowing what I know now, I'd have to say yes."[15]

Listening to all this, Joe Almeida thought the questions and answers were beside the point. The harm done to the boys in the Science Club had nothing to do with tiny doses of radiation. The key issue was the way he and the others had been confined by the state so they could be made available as research animals. The oatmeal experiment was just an example of a much larger scandal.

A single hearing, where politicians could promise reforms and witnesses were allowed to speak their minds, was not going to achieve the kind of justice Joe wanted. But he believed that more opportunities lay ahead. He planned to continue his own investigation, using whatever means necessary. He and Fred were beginning to think that it might be possible to sue someone—the state, the U.S. government, MIT, Quaker Oats—for damages. Maybe they could get some cash, or even an apology.[16]

Even though Sandra Marlow was gone, Joe Almeida continued to visit the Fernald library. He tried to keep tabs on Doe West, a former chaplain at the school who had been named to coordinate the state task force that was gathering the facts about the radiation experiments. Whenever Joe saw them, West and her small staff seemed overwhelmed by the amount of material they needed to review and the pressure they felt to complete their work. State officials were eager to announce that all the important facts were in and to put an end to press reporting on the scandal.[17]

To make sure that the task force had as much evidence as possible, Joe brought Fred to the campus and used his passkey to return to the attic of the Administration Building. Sandra Marlow went along, and the three of them rifled through box upon box. They collected more Atomic Energy Commission correspondence, and reports made by Dr. Benda. After sorting through it, they delivered several boxes of this material to the group at work in the library.

Although Marlow was no longer charged with developing the Fernald library, she was no less obsessed with the institution's past. In the middle of February, a friend showed her a newspaper ad announcing an estate sale at the home of Clemens E. Benda, whose widow had recently died. (Benda himself had died in 1975, during a trip to Germany. He was seventy-six.)[18] The estate's valuable art and furniture had already been auctioned, but the public was being invited to pick over what was left in his three-story stucco house in Arlington. She told Joe Almeida about the sale, and he agreed to come and to call Fred and bring him, too.

On a cold and cloudy Sunday, Marlow drove alone to the Arlington address, walked past wildly overgrown shrubs, climbed the steps to the front porch, and went inside, where a crowd of people examined old lamps, decorative items, books, and bric-a-brac. Fred and Joe arrived minutes later, but by prior agreement they didn't acknowledge Sandra. A woman who worked for the auction house told them that everything was for sale except for some papers stored in the attic. While Sandra pretended to look around downstairs, Fred and Joe went to find those papers.

Once they were in the attic, Joe and Fred opened files that held patient records, reports on various studies, autopsy findings, and memos to government agencies. Realizing that they couldn't just walk out with the stuff, they made a pile of books and then tucked the most tantalizing papers between their pages. One that held special interest for Fred noted the details of a study in which doctors at Fernald had pierced the sternums of patients to withdraw samples of bone marrow.[19]

When they had stuffed every available book and stacked them neatly, Fred and Joe began bringing them downstairs. They offered to buy the lot for a price that worked out to pennies per volume. The saleswoman agreed, and they loaded them into Fred's van.

Limited to what they could hide between the pages of books, Fred and Joe had left behind hundreds of pages of potentially valuable information. Early the next morning, they returned to take the plastic bags filled with trash that had been left outside the house for collection. They took them back to Fred's house, spilled them on his kitchen floor, and searched for papers that might relate to Fernald. They found nothing, but they satisfied themselves that they had done everything they could.

In the months that followed the sale at the Benda house, Fred and Joe would pore over their cache of papers and books and Sandra Marlow would do the same with materials she collected from university libraries and government sources. Though they cooperated, their trust in each other was not complete. Joe and Fred didn't share everything they had discovered with Sandra Marlow. She also kept a few items of her own research to herself.[20]

The former State Boys and the onetime Fernald librarian had some differing interests. Marlow was curious about medical research and government cover-ups. Fred and Joe were eager to find anything that documented the identities of Science Club members and the abuse and neglect they had experienced as boys. They preserved old internal reports of employees' misconduct and mysterious injuries suffered by Fernald residents. They were delighted to find a small book detailing every escape that occurred during the years they were at Fernald, because each of their names appeared in it many times. This book sparked memories of their adventures outside the Fernald gates and of their days and weeks in Ward 22.[21]

While Fred and Joe found proof of the injuries suffered by the State Boys, Sandra Marlow was able to put the oatmeal experiment in the larger context of controversial Cold War science. She discovered that in the 1950s Clemens Benda was a part of a group of physicians and scientists—the Cambridge Neurobiological and Psychedelic Study Group—who were deeply interested in the use of a recently discovered hallucinogen called lysergic acid diethylamide, or LSD. In 1956, Benda had joined Max Rinkel in a study that involved giving LSD to a subject so he would experience a psychotic episode. In a paper he published in 1957, Benda pointed out the most Strangelovian implications of this work.

> Let us imagine for a moment a situation where a master mind has used an agent like LSD on a large part of the population, and made them psychotic, without the knowledge of any contemporary. An investigating committee sets out to discover the cause of this psychotic epidemic. These investigators would not only need hundreds of cases in order to reach any significant conclusions, but also would need to examine these cases at a specific moment after the noxious agent had been liberated in the body. It is likely that a few minutes or hours later, only disintegrated metabolic products could be discovered in the organism, products which may have a metabolic level that occurs in the organism and, for that reason, seems to have no significance.[22]

With a paragraph that describes a mastermind and refers to people as cases and organisms, Benda showed an unseemly callousness for a man who promoted himself as a champion of vulnerable children. Sandra Marlow suspected something more sinister, given Benda's association with Rinkel. Like Benda, Rinkel had fled Germany prior to World War II. His pioneering research on LSD had been funded in part by the CIA, which had carried out secret studies with lethal results for some participants. (In one experiment, agents slipped LSD into the drinks of eight men. One subsequently fell to his death from a hotel window.) Marlow, who was devoted to exposing hidden elements of Cold War history, found Rinkel to be a thoroughly compelling figure.[23]

In the detritus of Benda's career, Marlow found papers on eugenics and even a photo of Nazi officers at a prewar medical meeting. (Benda was actually anti-Hitler and fled Germany to escape Nazism.) She was intrigued by a letter in which Benda showed an interest in "autopsy material" gathered at the Dachau concentration camp. And she was alarmed by a report on two human heads—collected during the Spanish Civil War—that were discovered to be stored at Fernald in the 1980s. It all seemed so darkly mysterious that Marlow felt a chilling sense of evil.[24]

But no evidence of a great atrocity emerged from the records. Benda had worked in Germany before the war, and while the sight of swastikas on uniforms was discomfiting, the pictures in the doctor's possession depicted what was a normal event in that place and time. Likewise, Benda's interest in human tissues may have seemed morbid, but it was routine for a neuropathologist. The heads, it turned out, belonged to a different scientist, who had left Fernald without taking them along.

Ultimately, the record suggested that Benda's most serious sin was grandiosity. His failure to obtain informed consent from Science Club participants was part of a pattern in which he placed enormous value on his own opinions. He had pressed on with his search for a hormone-related cause of Down syndrome even after his theory was eclipsed by the discovery of its genetic origins. And in retirement, he had felt it was his place to publish articles on subjects as widely divergent, and as far outside his field of expertise, as female orgasm, the mental health of Zelda Fitzgerald, and the dangers of putting fluoride in public water supplies. In the words of one Harvard professor who knew him, Benda was "a person with vigor, ambition and hustle. He could have been in used cars or suits; he happened to be in medicine. I don't think he was evil any more than a person who hustles Plymouths is evil."[25]

The Clemens E. Benda revealed in the files was ambitious, self-important, and well connected. But he was not singularly responsible for the Science Club experiments. And while he did nothing to improve their lot, he was not responsible for the suffering endured by the State Boys. Joe and Fred and Sandra Marlow slowly came to see that the boys who had been test subjects and state wards had been harmed by something bigger than any individual. They were victims of a way of thinking, or perhaps a dark part of human nature, that allows otherwise reasonable people to decide that a certain few are lesser beings.

*　　*　　*

Public revelations of Cold War radiation experiments continued well into the spring of 1994. Hundreds of cases were reported across the country, and the number of subjects rose into the thousands. The media's appetite for their stories was enormous. In March, Sandra Marlow and Fred Boyce went to New York to appear on a TV talk show. Not yet ready to be publicly identified, Fred appeared in silhouette, and the host of the show called him David. This time he was able to speak his mind about conditions at Fernald. He described crowded wards so cold the boys slept with their coats on, and a life so bleak that he leapt at the novelty of being in the Science Club.[26]

"But it was not for any medical purpose whatsoever," said "David." "And Fernald was just a nightmare. We were always praying and hoping that someone would come there and expose the place. . . ." He told the audience that he had actually fantasized that the visiting doctors would recognize that he didn't belong in an institution. "But no one would have ever believed us," he added, "because we were supposed to be retarded."

On the TV show, Fred said he feared that the state task force report on Fernald would be a whitewash. At the time, the panel members were busy reading documents forwarded to them by Doe West and meeting to discuss what they saw. Two camps emerged within the task force. On one side were state officials who wanted a quick report that skirted matters of liability. They didn't care to explore the relationships between scientists, state agencies, and Quaker Oats. And they tended to believe the experiments had been conducted in accordance with the standards of the times. On the other side were members of the task force who wanted more time to dig deeply into Fernald's past. They were less concerned about protecting those who had done the experiments and less inclined to shield them from liability claims. They also believed the Fernald experimenters had violated ethical standards that existed at the time of their work at the institution.

The more aggressive faction was led by David White-Lief, an attorney who had been chairman of a human rights committee at the Fernald School. Accustomed to the thorough style of investigation performed by litigators, White-Lief was disappointed by Reverend West's efforts. Working with little help, West had produced what he regarded as an introduction to the issues, not a complete examination.

In some ways, White-Lief would say, the task force had been rigged to produce limited results. The panel could not obtain subpoenas to require agencies and individuals to hand over documents or testify. It could only make requests, and in many instances these requests were ignored. (Some Boston area hospitals, for example, didn't respond at all to requests for information from the task force.) In the meantime, state officials pushed for the work to be finished. Though he made an effort to slow down the process, White-Lief would finish his service knowing that "we had left quite a few stones unturned."[27]

West's first draft of the task force report was so ambiguous and informal in its tone—no clear findings of fact were included—that the group requested a complete revision. When that was done, they haggled over the main points in a section titled "Findings and Recommendations." Although White-Lief lost a battle to have the final task force report declare that the Science Club experimenters had violated the Nuremberg Code, he did succeed in having other strongly worded items included. Among other points, the final report noted:

- Subjects and their families were not properly informed about the nature of the experiments.

- The studies had no potential benefit to the participants.

- Improper inducements had been offered to win the boys' cooperation.

- Scientists had exceeded government-set limits for radiation exposure in the experiment.

- State officials had failed to protect their wards from jeopardy.

These statements confirmed for Fred Boyce and the other State Boys that they had been wronged. But many of them took greater comfort from a paragraph that appeared in the preface of the report and was not included in press accounts. It established, in an official way, one of the most important facts about the State Boys and the system of institutions and ideas that had imprisoned them through childhood.

Please take note. Societal and cultural norms of the day permitted persons to be admitted to state institutions for a number of reasons. All were labeled mentally retarded just by virtue
of having lived within the facility . . . take special note that
these labels are grossly inaccurate, misleading, and simply not
true.

After describing what had gone wrong with the Science Club experiment, the task force recommended remedies. Most had to do with
changing laws to prevent future abuses. However, the panel did say
that the experiment subjects deserved compensation "for any and all
damage incurred as a result of such research."[28]

The damage done was a matter of civil rights and mental anguish. Just
who might pay was left unanswered. David White-Lief put the moral
burden on the state, Harvard University, and MIT. "The report calls for
the responsible people to step up to the plate, admit they violated people's
rights, and close this book once and for all," he told the press.

White-Lief recommended payments based on moral duty because
he feared that the Science Club members had little chance of getting
satisfaction any other way. The state appeared to have legal immunity
from lawsuits of the type they might bring, and the facts of the case did
not favor the study subjects. For example, they would have trouble
proving physical harm, because the radiation doses they had received
were low and the science of radiation and health rarely confirms that a
specific illness was caused by a specific exposure.[29]

As he considered the case, Sandra Marlow's lawyer friend Dan
Burnstein agreed with White-Lief. He believed that the members of
the Science Club would be stymied by the statute of limitations and
the state's legal immunity to most injury claims. The best remedy—an
apology and some payment—would come voluntarily, from officials
who concluded that it was the right thing to do. But as weeks, and
then months passed, no officials came forward.

TWELVE

"Okay?"

"All right."

"Uh, I'm Wally Cummins, the executive director of RADLAW, the radiation health effects public interest law group. We're here today to interview Fred Boyle."

"It's Boyce. B-O-Y-C-E."

"Right. Fred Boyce—B-O-Y-C-E—who was at the Walter Fernald School in Waverley, Massachusetts. Let's dive right in. Fred, how did you end up there?"

"Let's start by saying that when I first went there, at seven years old, the sign said THE WALTER E. FERNALD SCHOOL FOR THE FEEBLEMINDED. They had it on the sign right there. We had to deal with that."

In the video that documents his first meeting with one of the lawyers who would represent the Science Club, Fred explains that the promise of a party and escape from the routine of life in the BH had been more than enough inducement for the Science Club volunteers.

"We were told nothing whatsoever about radiation or anything that could be harmful. It was vitamins. And when we were told we would get extra privileges, and a party, we said, 'Wow! I want in!'"

With just a few anecdotes, Fred supplied Cummins with testimony that would make it impossible to believe that the Science Club participants knew what they were being fed with their oatmeal. Fred also spoke of additional experiments involving drugs and X-rays administered for no apparent reason. But he had trouble staying on these topics. Instead, he kept returning to his memories of life in the institution.

The State Boys, and the conditions at Fernald, were far more important to him than the oatmeal experiment.

A former law professor, the fifty-seven-year-old Cummins had been
nervous about meeting Boyce but managed to exude an air of extraor-
dinary calm. He delivered open-ended questions in a flat, emotionally
neutral voice. In response, Fred told highly dramatic stories of abusive
attendants, escape attempts, labor in the Fernald garden, and confinement in Ward 22. He told Cummins about getting Monopoly money
from the boy named LeBrun and about the time he was thrown into a
cell at the North Building. Fred spoke calmly until he began to
describe how the men in North Building lived in filth and were herded
like cattle. This memory forced him to pause so that he wouldn't cry.
"Forget about what they did to me," rasped Fred. "The way they
treated those men. And in *this* country."

The details Fred shared about the experiment and life at Fernald
made Cummins think that he might actually prevail in a suit against
those who had conducted and approved the experiment. At the very
least, Fred could shame the defendants into reaching a settlement. But
as much as he tried to concentrate on the legal implications of what he
was hearing, Cummins found himself dwelling on the emotional
power of Fred's stories. He couldn't imagine that Fred had been
deemed a moron and locked away. Even more astounding to him was
the notion that there were many more like him. He began to think that
the Science Club subjects had a claim to justice that reached far
beyond the oatmeal experiments.[1]

When he returned to Washington, Cummins began assembling legal
arguments and historical research to support a possible lawsuit against
the institutions responsible for the Science Club experiments. He
shared the work with his partner, Cooper Brown, who was a veteran of
desperate causes. Brown had worked for years with the National
Association of Atomic Veterans, which was organized in 1979 to pressure the government for information and, ultimately, compensation,
for men and women who had been exposed to radiation during atom
bomb tests.

Even those who were committed to the cause would have admitted that the atomic veterans, and others who alleged to be harmed by government science, suffered from serious credibility problems. As they spoke of projects with strange code names—Cal-2, Greenrun, Project 48-A—and ominous purposes, they could seem like paranoid conspiracy buffs. Who would believe that government researchers intentionally blanketed farm communities with radiation, or injected unsuspecting hospital patients with plutonium? Inside the Department of Energy, they were referred to as "the crazies," and their stories left reasonable people disbelieving.[2]

But in the 1990s, as documents had been declassified, it turned out that many of the crazies were right. Thousands of Americans had been secretly subjected to experiments, including radiation releases and plutonium injection. As the truth emerged, Brown and his partner Cummins were among the few lawyers in the country who had credibility with the victims of these experiments, because they had believed them when no one else would.

Beginning with a case in which civilians were injected with plutonium in Rochester, New York, Cummins and Brown began building files on behalf of many different groups of research subjects and their survivors. They got involved with families whose relatives were irradiated at a hospital in Cincinnati, and with similar plaintiffs in Washington State and Tennessee. In each case, Brown and Cummins would investigate, identify possible claims, and then find local partners to handle the actual litigation. These lawyers would file the complaints and motions and appear in court.[3]

Some of the most important evidence for the Fernald case emerged from the cardboard boxes of papers and files that Fred and Joe Almeida had filled after their various research expeditions. When Cummins returned to Massachusetts in the fall of 1994, Fred invited him to his house where he met Joe and they went through the documents page by page.

The first time he saw the men together, Cummins thought that Fred and Joe were Don Quixote and Sancho Panza come to life. Tall, thin, and determinedly optimistic, Fred was convinced of the basic goodness in people and certain that the State Boys of Fernald would find justice. He clung to these optimistic beliefs despite what had happened to him at Fernald, and despite the total silence that followed David White-Lief's call for officials to take responsibility for the Science Club debacle.

Joe, Fred's partner against the powerful, was less generous in his view of human nature. His anger about the past was more intense, and he had had a more difficult struggle to find stability in his adult life. But Joe was as loyal to Fred as a brother, and he was very good at finding the practical means to advance their cause. Cummins laughed out loud when he heard that Joe had returned to Fernald as an employee and gained access to its secrets.

Among the secrets that emerged from the cardboard boxes was the list of Science Club members. This was followed by Benda's file on the sternal puncture research. Last came a logbook showing injuries—many of them suspicious—to Fernald school residents. "The ledger of broken arms" is what Cummins came to call the book, and he believed it proved that the Science Club participants lived under a regime of terror that denied them any ability to make an informed decision about anything.[4]

Although the violence recorded in the ledger was emotionally powerful, the list of Science Club participants was much more valuable because it bore the names of thirty-five potential clients for lawsuits. In the winter of 1994, Fred and Joe would work the telephone and travel across New England looking for every man named on that sheet. Fred recalled them all as the boys who had lived with him in the wards of the BH and the BD.

A few of the State Boys had never moved more than a mile or two away from Fernald. Joe knew one who worked at a supermarket close to the campus. He had seen him many times, and they had traded stories about the past. But when Joe approached him to join the lawsuit, no amount of talk would persuade him. Fernald was in the past, he said, and even with the prospect of getting compensation, he wanted to leave it there. A handful of others had similar responses. One man had never told his wife and children about his childhood. Another was afraid that somehow the state would retaliate and he would wind up back in an institution.

Fear and self-doubt were epidemic among the men whom Fred and Joe managed to track down. In Waltham, Larry Nutt confessed that sometimes he felt that he had deserved to be placed in Fernald, even though he understood this wasn't true. Suffering from kidney disease and high blood pressure, Nutt railed about the loss of his childhood. He had never thought much about the Science Club. Like many of the others, he recalled that the experiment had been a blessed break from the routine at the school. But if a lawsuit meant getting back at the commonwealth, getting some restitution for those years filled with state-sponsored fear and abuse, he was all for it.

Not far from Nutt's house, Fred discovered Bobby Williams still living in the kind of boarding house that many State Boys had occupied during their first months on parole. Bobby worked menial jobs—including janitorial work at Fernald—to pay his rent and other expenses. He still missed his twin brother Richard, the one who had gone to Bridgewater after the Ward 22 riot. Whenever Fred saw him, Bobby talked about Richard's death with strange, detached precision.

"He had a weak mind, and I believe he was trying to kill himself when we were at Fernald and he drank that green paint," said Williams during one of these meetings. "Then he tried to kill himself again, in 1974, with a bottle of barbiturates. Then, on July 3, 1985, at 5:35 A.M., he died at Massachusetts General Hospital in Boston. He had taken a drink of hydrochloric acid. He really wanted to die that time, and the acid did it."

Richard Williams was the only suicide Fred discovered, but he was not the only State Boy who had died. Willie Adams, a dominant figure among the State Boys when they lived at Fernald, had grown into a hard-drinking man. He died alone, of alcohol-related disease, shortly after the Science Club story broke. Earl Badgett, one of the few African-Americans at Fernald, had spent roughly half his adult life in prison or jail on various drug-related charges, including possession of heroin. He had died of a heart attack while serving a sentence in a state penitentiary in Connecticut.[5]

Though Fred was disturbed by the circumstances in which the three men died, he was more impressed by how well some of the State Boys had done. About half were married with children. Many owned homes, and a few ran their own businesses. Jimmy Croteau, for example, had a thriving wallpapering business.

As happened with many of the State Boys, the moment Jimmy saw Fred, they fell into a very comfortable relationship. They were like long-separated siblings who were suddenly reunited. In one long night of conversation, Jimmy told Fred a story he had never shared with anyone else, a tale that included a loss unique among the State Boys.

As Jimmy explained it, in the months before his parole from Fernald he had begun a sexual relationship with a female attendant. Their affair continued after his release, and she soon became pregnant. He told Fred that state officials had threatened to return him to the school if he didn't sign an agreement ending the relationship and giving up his claim to the child. He signed, but had regretted it every day since.

All of the former State Boys told Fred that they had suffered for their lack of education. Some had managed to teach themselves basic skills in mathematics and reading. Others depended on help from wives, children, and friends. To a man, they confessed to having certain emotional problems. They found it difficult to trust others and to express themselves honestly. Nevertheless, many had managed to have long, loyal marriages.

Albert Gagne, for example, had married his wife, Doris, in 1964. A year later, they had a child, a girl they named Karen. Albert and Doris had just celebrated their thirtieth anniversary when they met with Fred to discuss the case. In some small ways, Albert still seemed like the State Boy whom Fred had known at Fernald. He was still shy, and he spoke slowly. Sometimes he had trouble finding the right word. But overall, Albert appeared to have overcome his past. He had found a long-term job at the Central Maine Medical Center, Lewiston's big hospital, where he won honors as a valued employee. He was a sober, churchgoing man who supported his family single-handedly and truly adored his wife and daughter.

Doris and Karen Gagne had been deeply affected by Albert's stories of growing up as a State Boy. Over the years, they had encouraged him to talk about his life and had gone with him to look at the Fernald campus. They would come to see the lawsuit as one more way for Albert to come to terms with his past and perhaps overcome it. Fred would encourage all of the men to look at the suit in this way. At some point, the state and others might be forced to pay the former State Boys money, but this was not guaranteed. The only payoff that Fred was sure about would come with the satisfaction of knowing that the truth was finally out.

Massachusetts officials had tried to satisfy the public demand for the truth with the report of the Fernald task force. When the work of the task force was completed in the spring of 1994, the Science Club members lost their forum and the local press moved on to other stories. However, the larger matter of America's radioactive history continued to occupy the Clinton administration in Washington. Hazel O'Leary, secretary of energy, had urged compensation for victims of federally sponsored research, and the president had created an Advisory Committee on Human Radiation Experiments to gather the facts and recommend a response.

In April 1994, the presidential committee began plowing through thousands of pages of documents, many of them recently declassified. This federal investigation was far more rigorous than the one carried out in Massachusetts by Doe West. A staff of more than forty researchers and analysts was hired to chase records. The committee members included internationally known experts in medical ethics, radiation and health, medicine, and law.

While the advisory committee staff collected thousands of documents, the public was invited to testify at more than a dozen hearings in six different cities. Within six months, the panel confirmed that more than 23,000 Americans had taken part in more than 1,400 different experiments. Officials acknowledged that the experiments posed legal and ethical problems and that as early as 1947 efforts were made to hide the radiation research to avoid scandal.

In mid-December 1994, Fred drove to Washington to testify before the committee, which met in a ballroom in a downtown hotel. For the first time, he was going to appear without the protection of anonymity, and this made him worry about speaking well, appearing intelligent, and being regarded as normal.

At the session Fred attended, the committee heard from relatives of people who had been subjects in what was called a "total body irradiation" experiment at a hospital in Cincinnati. Families of men who had been exposed to radiation at nuclear weapons factories in Ohio also testified. And in the most emotional moment of the day, the committee listened to a witness named Lenore Fenn describe an experimental treatment given to brain tumor patient Jacob Lifton at Massachusetts General Hospital in 1955. "Brain tumor and tissue spilled out onto the floor. Jacob wept, struggled, prayed and cried out," said Fenn, who had witnessed the procedure. "There was no treatment. This experiment was not for the benefit of the patient."

By the time he was called to speak, Fred Boyce had heard enough to understand that the committee would not be surprised by anything he had to say. He took a chair at the witness table and appeared to be perfectly calm. He began by placing the Science Club experiment in the context of life at Fernald. He described the isolation, the loss of autonomy, and the overcrowded wards. "Keep in mind we didn't commit any crimes," he said of the State Boys. "We were just seven-year-old orphans."

As the committee of lawyers, doctors, and scientists listened, Fred's voice wavered as he tried to connect the isolation and neglect of the State Boys with their role in the Science Club. They were state wards, he said, completely vulnerable to manipulation and exploitation.

> I won't tell you about the severe physical and mental abuse, but I can tell you it was no Boys Town. The idea of getting consent for experiments under these conditions was not only cruel but hypocritical. They bribed us by offering us special privileges, knowing that we had so little that we would do practically anything for attention. And to say, I quote, "This is their debt to society," end quote, as if we were worth no more than laboratory mice, is unforgivable.

Fred was so upset when he completed his testimony that his voice was tight, almost raspy. The crowd in the hall applauded when he was finished. He had intended to make the committee see that the State Boys had been victimized long before the Science Club came into being. He believed he had achieved his goal. But he would have to await the committee's report to discover if he was right.

As it examined records that covered five decades' worth of secret science, the federal panel made much of what it unearthed public. In doing so, the panel aided the lawyers who wanted to turn the Science Club subjects into plaintiffs in various lawsuits. Cummins and Brown gathered every document that the panel made available and fed their analyses of the material to their cocounsels around the nation.[6]

For the Science Club case, which would be filed in federal court in Boston, Cummins and Brown turned to an aggressive local firm called Dangel & Fine. In the early spring of 1995, one of the firm's lawyers, Michael Mattchen, took a Saturday drive down to Fred Boyce's house in Norwell to meet his would-be clients.

On that afternoon, the narrow street outside Fred's house was lined with cars bearing license plates from Maine and New Hampshire as well as Massachusetts. As Mattchen parked, he saw a few men standing outside smoking cigarettes. As he went inside, they flicked the butts into a snow bank and followed him.

With wives and children included, the group that greeted Mattchen in Fred's living room totaled more than twenty people. Fred asked him to sit on a folding chair he had set up next to his television. He then told the others to occupy the sofa, other chairs, and the floor. Then, one by one, Charles Dyer, Jimmy Croteau, Joe Almeida, and all the others stood to tell their stories.

A forty-six-year-old transplant from the Midwest, Mattchen was a soft-spoken man with blue eyes, prematurely white hair, and an extremely patient manner. Nothing in his face revealed to the men the outrage he felt as he heard about their lives at Fernald.

After all the men had spoken, Mattchen explained that he had come to Norwell with some reservations about the case. Given the low levels of radiation involved, he knew that it would be almost impossible to prove the Science Club had been harmed physically. And he knew that so much time had passed since the experiments that the defendants would argue that the statute of limitations had long run out. But after hearing the men talk, hearing the pain in their voices, Mattchen couldn't turn them down. Long after the afternoon had turned into a dark night, Mattchen waited for Fred to ask if anyone had anything more to say. When not one hand was raised, Mattchen stood up. He explained the problems with the low dose and the statute of limitations. He said the defendants would likely draw the battle out for years, and even then, he wasn't sure he could prevail. Finally, he said that, despite the odds, he wanted to represent them as their lawyer. After a moment of silence, Mattchen's message sunk in, and the men began to applaud.[7]

The Science Club suit would depend on the memories of those who participated, many of whom had tried hard to forget everything about Fernald. The legal team hired a Washington-based researcher named John Kelly to travel the Northeast to collect their stories. Few people could have been better prepared for this task. Born in 1941, Kelly was roughly the same age as the men he would interview. He had grown up in the Boston area, where he was educated in Catholic schools and infused with a religious devotion to the poor and the powerless. He held a master's degree in psychology and, more importantly, had worked at Bridgewater and Fernald in the 1960s. He had taken part in the early efforts to free many residents of institutions and had gone so far as to inflate test scores in order to get some of them released.

Kelly's experience at the state school gave him a vocabulary that made the men he interviewed feel more relaxed. It also meant he could skip a lot of basic questions about the campus and the routine at the institution. He also knew his way around the suburbs, small cities, and rural towns where the men had settled.

By the time Kelly began his interviews, many of the men had begun to worry that their health, and the health of their children, might be affected by the experiment. A few of them had no health insurance and suffered from chronic conditions like high blood pressure that went untreated. Some of the men reported suffering from lifelong psychiatric illness, and two were in such terrible shape that Kelly believed they needed immediate assistance.

> E.L., who looks like a Dachau victim, said he "always feels lousy." And has constant nightmares because of Fernald. He kept interjecting, "I'm afraid." He said he "couldn't go to the bathroom" and had low blood pressure.

"I was smart one time," E.L. said.

E.L. lives in a squalid, microscopic apartment with his wife Barbara, who is also a resident of Fernald. Both seem intelligent but have never been taught how to manage a home. The exposed mattress was filthy with black soot marks and clothes were in a pile on the floor and there was no room to walk or sit except on the cluttered bed.

E.L. . . . is in serious need of medical care but won't go to a Doctor because he can't afford to pay.

Many of Kelly's interview subjects were able to describe the oatmeal experiment in detail. A few recalled being told that they were getting special vitamins that would make them healthier and stronger. Every one of them spoke of the field trips and gifts they had received. For decades, the men had recalled these rewards with some pleasure. But with the truth of the experiment scandal, the gifts, banquets, and ball games were reinterpreted as part of a terrible betrayal. "I'd like to blow them away," said Larry Nutt, "but I don't know who they are."

As Kelly conducted dozens of interviews, he began to remember more of his own experiences at Fernald in the 1960s. He had worked in the behavior-modification lab under the leadership of a psychologist who had been trained by B. F. Skinner. He remembered giving psychological tests to children who scored high enough to be considered normal but had learned in their years of confinement to behave as if they were disabled.

"The main thing I learned there, which I remembered as I was seeing these guys, was that Draconian policies often start with the best intentions," said Kelly. "I mean, these guys had their lives ruined because people were trying to do good. That may be the scariest thing about it."[8]

On October 2, 1995, Fred was working a fair in North Carolina when he checked his home telephone-answering machine and heard a message from Cooper Brown. President Clinton was going to receive the federal advisory committee's report and issue the nation's apology to the victims of Cold War science. Fred had been invited to attend as a representative of the Science Club. The only catch was that the event would be held in Washington in less than twenty-four hours.

The carnival season had been good enough for Fred to hire a helper, who could handle the concession in his absence. He drove much of the night, stopping only for a nap in the back of his van, in order to reach Washington by morning. He stopped at a YMCA to shower and shave and change into his best suit. Then he walked to a card shop to buy a gift for President Clinton. He picked up a stuffed cat, a greeting card, and a fancy gift bag. Fred then went to meet Brown at his office. They drove together to the auditorium of the Old Executive Office Building, an imposing gray edifice next door to the White House.

Copies of the 925-page report were piled on a table in the auditorium, and Fred took one as he entered. He then sat patiently while members of the committee discussed their findings. Participants in the most egregious experiments, who had been given massive doses of radiation and suffered serious effects, were due both an apology and financial payment, concluded the task force. But where the radiation doses were low, such as the Fernald case, they advised that an apology and free medical monitoring were enough.

Though he was disappointed with the committee's approach to compensation, Fred could not have asked more from the president. "Those who led the government when these decisions were made are no longer here to take responsibility for what they did," said Clinton. "So today, on behalf of another generation of American citizens, the United States of America offers its sincere apology to those of our citizens who were subjected to these experiments, to their families, and to their communities."[9]

The apology could have only sounded better to Fred if it had addressed all the suffering of the State Boys who had spent so many years in Fernald as victims of circumstance, bureaucracy, and official abuse. This was not possible, he knew, and so he accepted what Clinton said as an attempt to right just one of the wrongs that had been done to him and his friends. Other men and women who had languished in institutions across the country would never receive even this much justice.

After he spoke, the president visited with the men and women who had attended the meeting as representatives of experiment subjects. Fred, who was the president's senior by five years, managed to shake his hand, and would forever recall how Clinton told him that his apology was heartfelt. Fred believed him. Before the president moved on, Fred handed him the bag with his gift and the card, which he had addressed to Socks, the Clinton family cat. Inside, he had written: "There was a cat named Miser who lived at Fernald. He lived there for years. I hope you have many more years in your home in Washington." The president accepted the gift bag, said thank you, but didn't pause to open it. Instead, he handed it to an aide.

Fred hoped that once the meeting was finished the reporters in the room might ask him for an interview. But just as the press seemed to turn toward the subjects of the experiments, someone in the crowd announced that one of the biggest news stories of the decade—O. J. Simpson's murder trial—had ended with a verdict of not guilty. Every reporter but one fled the room. In the days that followed, their reports on the president's apology were buried by the coverage of the Simpson story.

One reporter, a young man who worked for National Public Radio, lingered long enough to interview Fred. That night, amid the avalanche of O. J. Simpson news, the story of Clinton's apology and Fred's positive reaction to the president's words were broadcast briefly on the NPR network. In Oklahoma City, Abra Glenn-Allen just happened to be listening to the broadcast in her car. When she got home, she called his home in Norwell and left a message of congratulations on the answering machine.

The president's statement was a moral victory, but the men of the Science Club wanted satisfaction from those who had been responsible for the experiment. They wanted to hurt the state, MIT, and Quaker Oats, and the only way to do that, they thought, was to force them to pay money.

Two weeks after Clinton's speech, the group that had rallied around Fred was surprised to read in the paper about a lawsuit filed by one of their old mates at Fernald. Ronald Beaulieu had gotten his own attorney, and they had quietly filed claims against the federal government, the state, MIT, and Quaker Oats. Though Beaulieu's lawyer, Jeffrey Petrucelly, got to the courthouse first, the filing had no effect on Fred's group. When their lawsuit was finally filed in December 1995, it was consolidated with Beaulieu's. Petrucelly and Mattchen would work together for all of those named in the suits, and for anyone who might have been a subject in the experiment but hadn't come forward. They were seeking $4 million per man.[10]

Though they could have been competitors, Mattchen and Petrucelly were comfortable as a team. Both men had come of age in the 1960s and had carried some of the political attitudes of that era into adulthood. They were wary of government power and enjoyed fighting for underdogs.

In the early stages of the case, Mattchen and Petrucelly put much of their effort into identifying clients. The question of just who was in the Science Club would prove difficult to answer. (The list Fred and Joe discovered didn't cover all the experiments.) Citing privacy laws, state officials would not publish an official roster of all Science Club participants, but they invited men who thought they were in the club to submit their names for review. Jimmy Croteau, who had vivid memories of the experiment, was told he had never been involved. Charles Hatch got the same answer when he inquired, and never even bothered to contact Mattchen or Petrucelly. However, other men who wrote to the state were recognized as participants in the experiment. The number of confirmed Science Club members who joined the suits eventually exceeded seventy. Each of these men would eventually receive from the state an accounting of his activities as a research subject. Albert Gagne would be identified as "Subject No. 29" in Experiment No. 10, which was conducted in July 1951. Along with seven other subjects, he ate oatmeal with radioactive calcium added. A similar number of boys in this experiment got radioactive farina.[11]

The federal judge assigned to the claims urged the sides to work with a mediator, retired judge David Mazzone. Knowing that New England juries can be stingy with awards, Mattchen and Petrucelly agreed to mediation. After they met with opposing lawyers, they quickly concluded that both MIT and Quaker Oats were eager to resolve the suits, but the state and federal governments seemed more inclined to fight. This may have been because they had less to fear when it came to negative publicity. And surely the government lawyers were alert to avoid the long-term consequences of any precedents that might be established by the suits. State and federal agencies enjoy certain immunities from claims, and they would not want to give any of these protections away.

The hearings conducted by Judge Mazzone were not as formal as a trial, but he did call on the lawyers to make their cases with vigor. The defendants raised the statute of limitations, argued that the radiation doses were innocuous, and then suggested that the era constituted a moment of national crisis that justified secrecy. To make their point, the federal lawyers held up an old newspaper with the headline JAPAN BOMBS PEARL HARBOR.

When they got their chance, Petrucelly and Mattchen knocked down the statute-of-limitations argument in a convincing way. The statute takes effect only after a person understands he was harmed, they noted, and since the state had kept the details of the experiment secret, the clock didn't start to run until the 1993 news reports on the experiment. The danger posed by the radiation itself was not part of the Science Club's claim, continued the lawyers. They were not arguing that they had been physically harmed. Instead, they insisted that their civil rights had been violated by the deceit of the scientists, state officials, and Fernald's doctors.

To emphasize the point that the boys at Fernald had been harmed, Mattchen turned to a prop of his own, a twelve-inch-long stainless-steel syringe and needle he had borrowed from a surgeon. A larger needle than this had been used to inject radionuclides into those boys who had received intravenous doses, noted Mattchen. Given that none of them had provided proper consent, a syringe full of radioactive solution administered with such a tool might well be considered battery on the part of the state.

Watching the arguments were three members of the Science Club—Robert Gagne, Gordon Shattuck, and Ronald Beaulieu—who had been summoned by Judge Mazzone to represent the larger group. After the lawyers sat down, the judge invited them into his chambers for a private meeting where, Gagne would recall, he told them that they had a legitimate claim. However, he also cautioned them that a lawsuit might take years to resolve, and even then they might lose.[12]

With the judge pushing both sides, it became clear that the defendants were willing to pay something—not $4 million per man, but something—to make the claims go away. In the end, Quaker Oats and MIT agreed to put up $1.85 million to satisfy the entire class of Science Club members. After a longer struggle, the Commonwealth of Massachusetts and the federal government together offered about $1.2 million. A formula was devised to determine how much each man would receive. Those who were married got extra for the distress suffered by the wives, and those who had participated in developing the suit from the start got something for their work. A typical payment would be between $50,000 and $65,000.

Although the money would hardly make a man rich, it would be a potent symbol. But Mattchen and Petrucelly wanted something more. They wanted formal apologies. Here the state, MIT, and Quaker Oats couldn't match President Clinton's generosity of spirit. Lawyers for the state said they wouldn't go beyond a letter in which Fernald's current director offered his "personal apologies for this unfortunate situation." The trouble was, the letter was so vague that one couldn't tell whether the "situation" was the experiment, or the uproar that accompanied the news that it had been done. Similarly, MIT declined to apologize because its president had already issued a fuzzy statement saying he was "sorry to hear" that informed consent had not been obtained. Quaker Oats refused to address the matter at all.

The defendants' lawyers may have had legal reasons for denying the Science Club members a clear, unambiguous apology. Years after the settlement, they would decline to talk about it. But their refusal didn't dampen the satisfaction the Science Club members felt after they met in a hall in downtown Boston and voted to approve the deal. To them, the money was an acknowledgment that they had been violated and that the powerful who had once ruled their lives had been forced to admit they were wrong. The defendants' refusal to add a candid apology to the settlement simply confirmed what they had always believed about big bureaucracies and the people who run them.[13]

AFTERWORD

A handful of Science Club members declined to take part in the law-suits and the subsequent settlements. Jimmy Croteau, for one, attended the first meeting held at Fred Boyce's house but wouldn't par-ticipate any further. He feared that too many people, including his daughter, would learn things about his past that he preferred to hide. But he did embrace his new friendship with Fred and came to depend on him more in late 1999, when he was diagnosed with cancer.

When Fred came off the road that year, he visited Croteau, who was staying with his ex-wife Priscilla Sneider. Though they had been divorced for decades, the two remained very close, and Jimmy relied on her to see him through the end of his life. The last time Fred saw him, Jimmy was bedridden and barely able to speak. He said he was afraid.

"I just sat by his bed and talked a little bit," recalled Fred. "I squeezed his hand and told him that he was just going to a new place, that it was okay to be scared but it was going to be all right. I said it was like when we moved from BD up to BH. It was scary, but we got used to it and it wound up being a better place."

After Jimmy's death, Priscilla began to discover the facts of her for-mer husband's life. She was saddened to realize that, except for his work, he had spent his last twenty years in almost total isolation. He had stockpiled food and toiletries as if he was expecting a disaster, and was so secretive that none of his neighbors knew him at all. Few could recall ever even speaking to him.

The sadness Priscilla felt as she and her daughter emptied his apart-ment was multiplied when Fred told her about the child that Jimmy had fathered but never met. No one in Jimmy's life had been more generous and reliable than Priscilla. Those qualities alone should have allowed him to trust her. But she had an even greater reason to feel upset over the way he had held this secret from her.

As it turned out, Priscilla's own aunt had also been a resident of Fernald in the 1940s and 1950s. She, too, had had an affair with a Fernald employee. She had become pregnant and was forced to give up the baby for adoption. Jimmy Croteau had known this story, and Priscilla couldn't believe that with this knowledge he still kept his secret. Had she known that Jimmy was pining for a child from the past, she would have done everything she could to unite them. He never gave her the chance.

Jimmy Croteau represented a tragic extreme among the State Boys who came together when the Science Club scandal erupted. The majority did receive money from the settlements and put the cash to good use. Albert Gagne saved most of his for retirement. Larry Nutt used his windfall to throw a big wedding for one of his daughters. Joe Almeida, who still worked at Fernald, bought a small house in Nashua, New Hampshire, and paid off some long-standing debts. The house satisfied one of Joe's lifelong dreams. He moved in with his children from his second marriage, Jason and Mandy. They were soon joined by Mandy's first child, a daughter named Miah.

Unlike Joe, Fred Boyce already owned a home, and he had no big debts. He expected to just put his windfall into the bank for the long-term future. Then, in the winter of 2000, he began to feel run-down and his stomach hurt. He had no health insurance, so he was reluctant to see a doctor. Finally, in the fall of 2001, when he could no longer ignore the symptoms, he stopped at a hospital emergency room in West Virginia. Doctors there made the diagnosis of advanced colon cancer.

In his moment of crisis, Fred called Abra. She insisted he come to Oklahoma. He did, and Abra's extended family—including her mother and a brother-in-law who is a physician—arranged for both shelter and affordable medical care. Surgery and chemotherapy occupied Fred through a desperate winter, but they produced a cure. He didn't stay to complete all of the recommended treatments, however. The carnival business, and his worries about money, called him to the road about halfway through.

By the spring of 2002, Fred was back on the carnival circuit. He was in Virginia in early May 2002, when then governor Mark R. Warner issued a formal apology for his state's eugenics law and practices. The occasion was the seventy-fifth anniversary of the Supreme Court decision that led to Carrie Buck's sterilization. By the time the practice ended in 1979, more than 7,000 Virginians had been sterilized. "The eugenics movement was a shameful effort in which state government never should have been involved," said Warner. The day before the governor spoke, two state lawmakers presented a commendation for World War II service to a Lynchburg man who was sterilized at age sixteen because he was a runaway.[1]

In 2003, two years after he was diagnosed, Fred was still healthy, and the cancer was becoming part of the life story he told when schoolteachers, who had read of the Science Club, invited him to speak to their classes. More than cash or the president's apology, these public speaking dates were Fred's big reward for his struggle. Just as he had hoped, the notoriety of the lawsuits had made it possible for him to tell the bigger story of the State Boys. Some teachers, such as Carol Kilpatrick of Mt. Vernon, Oregon, invited him year after year. In December of 2002, she videotaped her high school students as they gathered around a speakerphone to hear Fred tell his story over a long-distance line.

In the long question-and-answer session that followed his talk, Fred told the class about his life in foster homes, the conditions at Fernald, and his struggle to adapt to life on the outside following his parole. The students were so engrossed by Fred's story that they kept asking questions and taking notes even after the period ended and it was time for them to move to other classes. The next day, when teacher Kirkpatrick asked for their reflections on the talk, one young woman named Jasmine dwelled on how Fred's life showed that "one person could make a difference." Lisa, another student, said, "He wasn't bitter, even though everybody wronged him . . . that blew my mind."[2]

It is not likely that the students in Kirkpatrick's classes understood, completely, the difference between the times in which they live and Fred's school years. Today in the streets of big cities and the backwoods of rural America, one can find many older men and women who struggle just to survive in part because their learning difficulties were never addressed and they were deprived of an education and the access to jobs and social contacts that come with high school and college diplomas. Many are former State Boys and State Girls who never recovered from their experiences as inmates of institutions. In contrast, there were students at the Mt. Vernon school with learning problems serious enough to have landed them in a place like Fernald fifty years ago. Instead, as youngsters, they had been evaluated by experts and given special types of instruction, which will be available to them all the way through college.

The facts about learning disabilities are so widely known that it's not unusual for a boy or girl to be enrolled in a school program even before age five. Newspapers and television often present accounts of these children succeeding, and thriving, in regular school. In a typical story published in the *Boston Globe* in 2002, writer Barbara Meltz described the experience of a boy named Danny. Identifying his problem was difficult, and the boy struggled. Then, in second grade, he began a program of special instruction that he attended just four hours per week. Within months, he had caught up to, and passed, the reading level of his classmates.[3]

While Danny turned out to be an exceptionally successful student, he is not unique. Federal and local laws require special services for every child determined to have a learning disability, and the vast majority receive them. Similar laws keep children with mild and moderate retardation in local schools and out of institutions. Given these developments, it's certain nearly all of the State Boys and Home Boys that Fred Boyce lived with in the wards of Fernald would today be in regular schools, even if they were in foster care.

Fred Boyce insisted he was not bitter over his fate, but he and other former State Boys never escaped their memories of life at Fernald. In early 2003, just weeks before he died of heart failure, Larry Nutt sat in the kitchen of his house in Waltham and cursed the system of tests, institutions, and experts "who ruined my life. I don't believe we know all the stuff they did to us," he added. "I don't know if we'll ever know the whole truth."

Shortly before Nutt died, Harvard University opened the collection of papers it had received from the estate of Clemens E. Benda. These records, and others housed in state archives, added to the picture that Larry Nutt knew was still incomplete. Among the new materials was evidence of lobotomies performed on Fernald patients, a practice not previously made public. The records also identified a new group of previously unnamed Science Club members, including Charles Hatch, whose claims were rejected by the state in 1995.[4]

Three more previously unnoticed subjects were identified in Benda's papers as "spastic patients" who received between 2 and 4.6 microcuries of radio calcium. One of these participants, referred to as LF, was a boy of twelve with a reported IQ of 49. A second, FJD, was aged twenty-two and had an IQ of 61. The third, MN, was twenty-one, with the exact same IQ score.[5]

In one of the more curious entries in Benda's records on the oatmeal project, researchers noted that after twelve different feedings the results were identical to those of experiments done with puppies. After this fact was noted, they doubled the amount of radiation that would be fed to the boys.[6]

The Benda files also showed that besides radiation studies, Fernald scientists had used residents to test everything from psychosurgery to testosterone. The use of institutionalized children as research subjects was not confined to Fernald. For example, in the 1960s children at New York's Willowbrook School were given hepatitis by researchers who reasoned that they were bound to get it anyway. But Fernald, with its connections to the Boston medical community, was likely the most active center of research on institutionalized children in all of the United States.[7]

The school remained a home of last resort for the severely disabled and celebrated 150 years of operation in 1998. Though Fernald's population had shrunk to what was considered a manageable size—385—the commonwealth still couldn't provide reliable staff and safe conditions.

Two years before the 150th anniversary, government investigators issued a scathing report on problems at the Greene Building, which housed 80 people. As the *Boston Globe* reported, in just six years, the residents of Greene had suffered 1,400 injuries, including one violent rape in which staff were suspects but no charges were ever filed. In 119 cases, women suffered bruises indicating possible sexual assault. Auditors also reported visiting wards where outside doors were open and unguarded and every staff member was asleep. One attendant snored loudly and never awakened during the inspection.[8]

If officials could find assaults, abuse, and neglect at Fernald in 1996, where staff and programs had supposedly been improved, the stories told by the State Boys had to gain credibility. And revelations of once-hidden history in other parts of the country lent credence to the notion that the Massachusetts group was representative of a much larger population. In the year 2002, for example, the *Portland Oregonian* published details of a state program that forced the surgical sterilization of 2,650 people who had been diagnosed as disabled or mentally ill. The program had operated into the 1980s. In 2003, five people who had lived in an institution in Iowa in 1939 sued the state because it had allowed a researcher to condition them to stutter, even though they spoke normally, as part of an experiment.[9]

Fred Boyce and Joseph Almeida would have to wait for the satisfaction of seeing the dorms at the Fernald School knocked down by a wrecking ball. Although Massachusetts Governor Mitt Romney remained committed to closing the place, in the summer of 2004 he was delayed when advocates for some of the remaining residents went to court to stop the state from acting on its plans.

Among other things, those who opposed the closing feared that the last residents to leave Fernald—two hundred seventy-five in all—would be sent to worse places. To support their charge that the commonwealth was not prepared to offer quality care, relatives of Fernald patients filed affidavits charging that patients at public facilities still lived in filthy conditions, suffered physical and verbal abusive, and were dying at unexpectedly high rates.

Officials denied the essential argument posed by Fernald families—that the commonwealth could not or would not provide quality care—but promised to investigate the allegations. The specific charges were open to interpretation, but no one could deny that the latest development in the school's history held more than a little irony. After decades spent fighting for something better than Fernald, the relatives of patients were demanding it stay open because they feared the alternatives offered by the government.

Although the battle over the way Fernald was to be closed mattered to these it served and the community of Waltham, Boyce, Almeida, and the other state boys were focused on developments that affected them more closely. In May of 2004, they filed a formal request for an executive order that would update their state records to indicate they were not, and never had been, morons. The petition, drafted by attorney David White-Lief, also sought a formal apology for the way the men had been treated and the creation of a commission to determine payments for the labor that residents performed for the state while living at Fernald and other institutions.

While White-Lief and several state boys met with the commissioner of retardation to work on their requests, another Boston lawyer went to court to win satisfaction for Charles Hatch. As a seven year-old, Charlie had been coached by his stepmother to tell lies that resulted in him being locked away at Fernald. Though he says he had vivid memories of participating in the Science Club experiments, officials prevented him from receiving money in the settlement, claiming there was no record of his involvement. Under pressure from lawyer Jeffrey Petrucelly, the commonwealth finally admitted he had been used as a human guinea pig. When he filed suit to demand that Hatch be compensated, Petrucelly noted "there are many more" with similar valid claims.

Besides the legal maneuverings of families and former state boys, Fernald was the subject of one other initiative. With the support of Waltham's newspaper, *The Daily News Tribune*, Fred Boyce began a campaign for a community center to help all Fernald alumni get social services, education, employment counseling, and access to their health records. This place, which Fred called "a clubhouse" could be established and funded indefinitely with a small portion of the money Massachusetts will one day realize from the sale of the Fernald property.

The clubhouse proposal, and all of the legal initiatives, brought many of the state boys together and allowed them to renew their friendships. The attention these developments received also brought Fred Boyce's brother George back into his life. George appeared at a public forum in Waltham where Fred was participating in a discussion of the history of Fernald. In an emotional address, he challenged the idea that his mother was an alcoholic—"she just had too many kids"—and then described his shock at learning what had happened to his brother inside the institution.

Outside the meeting hall, Fred and George stood beneath a street-light and argued about their mother and other bits of history they recalled in different ways. The conversation was continued at one of Fred's favorite spots, Dunkin' Donuts, and ended with them agreeing to renew their relationship.

NOTES

FOREWORD

1. The story of Howie's punishment was related in separate interviews with three of those who were present: Frederick Boyce, Joseph Almeida, and Albert Gagne.

2. A thorough account of the treatment of so-called defectives is found in James W. Trent, Jr., *Inventing the Feeble Mind* (Berkeley: University of California Press, 1995). For the Florida case, see Edward Shorter, *The Kennedy Family and the Story of Mental Retardation* (Philadelphia: Temple University Press, 2000), 25.

3. From Walter E. Fernald, *History of the Treatment of the Feeble-Minded* (Boston: Geo. Ellis Co., 1912). See also Dr. Anna M. Wallace, *History of the Walter E. Fernald State School* (Waltham, Mass.: WEFSS, 1948); and Estelle Foote, M.D., *Walter E. Fernald: His Writings and His Clinics* (Waltham, Mass.: WEFSS, 1950). For a report on the population of institutions and an estimate of the proportion of those without mental defect, see Robert Wallace, "A Lifetime Thrown Away by a Mistake 50 Years Ago," and "Mental Homes Wrongly Held Thousands Like Mayo Buckner," *Life* magazine, March 24, 1958.

4. Wendy Kline, *Building a Better Race* (Berkeley: University of California Press, 1994), 20–21.

5. R. L. Dugdale, *The Jukes: A Study in Crime, Pauperism, Disease, and Heredity* (New York: G. P. Putnam's Sons, 1877). For an effective analysis of Dugdale, see Stephen J. Gould, *The Mismeasure of Man* (New York: Norton & Co., 1981). For a definition of eugenics and Galton's contribution, see Elof Axel Carlson, *The Unfit: Definition of a Bad Idea* (Cold Spring Harbor, N.Y.: Cold Spring Harbor Laboratory Press, 2001), 9–11, 143–46, 232–34. For Roosevelt and Progressivism, see Kline, *Building a Better Race*, 1–2, 11–14, 17–20.

6. Madison Grant, *The Passing of the Great Race* (New York: Charles Scribner & Sons, 1916).

7. Martin S. Pernick, *The Black Stork: Eugenics and the Death of "Defective" Babies in American Medicine and Motion Pictures Since 1915* (New York: Oxford University Press, 1999), 3–8.

8. Ibid. Photo from the State Historical Society of Wisconsin.

9. From Fernald, *History,* and Walter E. Fernald, "Thirty Years' Progress in the Care of the Feeble-Minded," Proceedings of the Forty-Eighth Annual Session of the American Association for the Study of the Feebleminded, May-June 1924.

10. Kline, *Building a Better Race,* 21–22.

11. Lewis Terman et al., *The Measurement of Intelligence* (Boston: Houghton Mifflin, 1916). More on Terman is in Mitchell Leslie, "The Vexing Legacy of Lewis Terman," *Stanford* magazine, July-August 2000, and in Richard C. Paddock, "The Secret IQ Diaries," *Los Angeles Times,* July 30, 1995.

12. Terman et al., *The Measurement of Intelligence.*

13. Henry Goddard, *The Kallikak Family: A Study in the Heredity of Feeble-mindedness* (New York: Macmillan, 1912). The criticism of this work is in Gould, *The Mismeasure of Man.*

14. Goddard, *The Kallikak Family.* For the story of the word *moronia,* see Kline, *Building a Better Race,* 23–26.

15. Henry Goddard, *The Menace of Mental Deficiency from the Standpoint of Heredity* (Vineland, N.J.: Vineland New Jersey Training School, 1915). The menace of the feebleminded is discussed at length in James W. Trent, Jr., *Inventing the Feeble Mind* (Berkeley: University of California Press, 1995).

16. Goddard, *The Menace of Mental Deficiency from the Standpoint of Heredity.*

17. Estelle Foote, M.D., "The Traveling School Clinic of the Walter E. Fernald State School," address to the 1951 conference of the American Association of Mental Deficiency, May 1951. See also Trent, *Inventing the Feeble Mind,* 151–55.

18. Foote, *Walter E. Fernald.* See also *Bulletin of the Vineland Training School* (Vineland, N.J., 1913).

19. Clifton Perkins, "A Medico-Civic Concept of Defective Delinquents and Psychopaths," speech delivered January 14, 1946.

20. Kline, *Building a Better Race,* 50–51.

21. The story of the Buck case is told in great detail in David Smith, *The Sterilization of Carrie Buck* (Croton-on-Hudson, N.Y.: New Horizons Press, 1989).

22. Records of state sterilization totals from Dolan DNA Learning Center, Cold Spring Harbor Laboratory, Cold Spring Harbor, N.Y.

23. Carlson, *The Unfit,* 344–49.

24. Harold Skeels and Harold Dye, *Journal of Psycho-asthenics,* Annual Proceedings of the American Association of Mental Deficiency, May 1939, 136.

25. Fernald and Terman's recantations are discussed in Gould, *The Mismeasure of Man.* Fernald's views are recorded in Foote, *Walter E. Fernald.* For Hoover, see Kline, *Building a Better Race,* 100–106.

26. The story of the Eugenics Records Office is told throughout Stefan Kühl, *The Nazi Connection* (New York: Oxford University Press, 1994). See also Trent, *Inventing the Feeble Mind,* 172–73, and the archives of Cold Spring Harbor Laboratory, Cold Spring Harbor, N.Y.

27. Kühl, *The Nazi Connection,* 59–63. See also "Yale Study, U.S. Eugenics Paralleled in Nazi Germany," *Chicago Tribune,* February 15, 2000.

28. Carlson, *The Unfit*, 327.

29. Kühl, *The Nazi Connection*, 99, 100.

30. Clarence J. Gamble, "The Prevention of Mental Deficiency by Sterilization," *American Journal of Mental Deficiency,* July 1951.

31. Admission notes on scores of children with relatively high IQs can be found in the Walter E. Fernald State School superintendent's notes, 1940 through 1960.

32. Trent, *Inventing the Feeble Mind,* 233–45.

33. Massachusetts figures found in *Massachusetts Needs in Mental Health and the Care of the Retarded, Commonwealth of Massachusetts 1958.* Population figures and number of institutions from U.S. Census. See also Richard McKenzie, *Rethinking Orphanages for the 21st Century* (Thousand Oaks, Calif.: Sage Publications, 1999). Proportion of normal and near-normal children at Fernald reported annually by school's psychology department.

ONE

1. The story of Frederick Boyce's life on the Bond farm is based on his recollections, and interviews with others, including Mrs. Bond's neighbor George Hoyt and former foster children who had also lived on the farm. Records of Commonwealth of Massachusetts child welfare agencies and Frederick Boyce's various state files were also used to confirm the account.

2. From social worker case files on Frederick Boyce.

3. Interview with Frederick Boyce, February 15, 2000.

4. IQ test questions based on the version of the Stanford-Binet Test used at Wrentham in 1947.

5. Interview with Frederick Boyce, February 15, 2000.

6. Psychiatrist's letter to social service agency.

7. Interview with Frederick Boyce, February 15, 2000, also supported by social worker case files.

8. Interview with Frederick Boyce, also supported by social worker case files.

9. Recollections of Mrs. Bond based on interviews with Frederick Boyce and her neighbor George Hoyt.

10. From social worker report in file of Frederick Boyce.

11. Ibid.

12. Mrs. Bond's death confirmed by records of Merrimac town clerk. Account also based on interview with Frederick Boyce and social service records in his state file.

13. Details of Boyce's commitment based on his social service case files, and interviews with him.

TWO

1. Details on the Fernald campus and buildings are based on tours, historical photos, engineering reports, and interviews with former staff and residents. Much

information was gained from William Gabler, *Report of Inspection of Walter E. Fernald State School* (Boston: Commonwealth of Massachusetts, 1955).

2. Jacques's name appears on Boyce's admission forms. Details of his arrival at Fernald are based on interviews with Boyce and those who were in the ward when he arrived.

3. The routine of life on the ward was recalled in repeated interviews with multiple sources including audiotape of former attendant Maude Bell held by researcher Sandra Marlow. Boyce's own experience is based on his recollection and on supporting details in his Fernald records.

4. Blacksmith's demeanor and other events based on reports of Robert Gagne, Albert Gagne, and Frederick Boyce. Interviews conducted throughout December 2002.

5. Mealtime and other patterns of life on the ward recalled by former attendant Sumner Noble as well as many former Fernald residents.

6. Edouard Seguin's work is described in James W. Trent, Jr., *Inventing the Feeble Mind* (Berkeley: University of California Press, 1995), 46–52. The application of Seguin's methods at the Walter E. Fernald State School is described in teacher education materials held by the school's library.

7. IQ scores reported in Boyce's Fernald records.

8. Farrell was notified in psychology department notes, December 1947, Walter E. Fernald State School. Stronger warnings about IQ tests were aired in Beth L. Wellman and Boyd R. McCandless, *Factors Associated with Binet IQ Changes of Preschool Children*, Psychological Monographs (Washington, D.C.: American Psychological Association, 1946).

9. The Gagne boys, their family, and their commitment are described in their Fernald records and social service department reports. The events described are also based on interviews of Robert and Albert Gagne, their sister Doris (Gagne) Perugini, and others.

10. Company Sundays are described in decades of Fernald School reports and by former residents and attendants. The specific incidents were recounted in interviews with those involved, December 2002.

11. From Joseph Almeida's social service and Fernald School records. Also based on interviews with multiple subjects.

12. From Robert Williams's state records, and interviews, November 2002.

13. McGinn and his actions were recalled, in detail, by more than a dozen sources, including former physicians, teachers, attendants, and residents of the Walter E. Fernald State School.

14. Ibid. Also, records of firing and hiring at Fernald are included with monthly superintendent reports.

15. Interview with Sumner Noble, January 12, 2002 .

THREE

1. From interviews with Robert Gagne and Kenneth Bilodeau, June 2002. For McLean Hospital, see Alex Beam, *Gracefully Insane: The Rise and Fall of America's Premier Mental Hospital* (New York: Public Affairs, 2001).

2. Life on ward from multiple interviews with Frederick Boyce, Robert Gagne, Albert Gagne, Joseph Almeida, and Charles Dyer, 2002–3. Descriptions of Farrell and his attitudes are from his contemporaries on the Fernald staff, including Kenneth Bilodeau, Rose Terry, and Pat O'Callaghan. "Items" is from Burton Blatt, *Christmas in Purgatory* (Boston: Allyn & Bacon, 1966), 89. Interviews conducted throughout 2002.

3. From curriculum vitae of Malcolm Farrell.

4. Federal funding for research discussed in Edward Shorter, *The Kennedy Family and the Story of Mental Retardation* (Philadelphia: Temple University Press, 2000), 91–94. Details on the types of research done can be seen throughout Marvin Rosen, Gerald Clark, and Marvin Kivitz, *The History of Mental Retardation* (Baltimore: University Park Press, 1976).

5. Reported in various statements to the Walter E. Fernald State School Board of Trustees, 1945 to 1960.

6. Benda's background is reported in *Who's Who in the American East.* His scientific interests can be seen in *The Thyroid Studies: A Follow-up Report on the Use of Radioactive Materials in Human Research Involving Residents of the State Operated Facilities of the Commonwealth of Massachusetts from 1943 to 1973* (Boston: Massachusetts Department of Mental Retardation, 1994). Benda's other interests, and brain samples, are described in Clemens Benda, *Research Projects at the Walter E. Fernald State School,* report to trustees, April 9, 1962.

7. "Nightmares and Tantrums Show Effects of Bombing on Children," *Boston Globe,* October 1, 1949.

8. Benda's travels are reported in his monthly statements to trustees. For Freeman, see Jack El-Hai, "The Lobotomist," *Washington Post,* February 4, 2001. For Rosemary Kennedy, see Shorter, *The Kennedy Family,* 91–94.

9. Benda, *Research Projects,* April 9, 1962.

10. Edward Hooper, *The River* (New York: Little, Brown, 1999), 201–2, 412–22. For overview, see also John R. Paul, *The History of Poliomyelitis* (New Haven: Yale University Press, 1971).

11. Farrell letter dated November 2, 1949.

12. Based on multiple interviews with Frederick Boyce, Charles Hatch, Robert Gagne, and Joseph Almeida conducted throughout 2002. For details of experiment, see *Report on the Use of Radioactive Materials in Human Research Involving Residents of the State Operated Facilities of the Commonwealth of Massachusetts from 1943 to 1973* (Boston: Massachusetts Department of Mental Retardation, 1993). See also *Clinical Director Reports to Superintendent,* Walter E. Fernald State School, 1949–55.

13. Based on interviews.

14. Interviews with Bea Katz, Louis Frankowski, and Frederick Boyce conducted throughout summer 2002.

15. From reports of farm production, held by Howe Library, Walter E. Fernald State School.

16. Interviews with Katz and Boyce.

17. Bernadine Schmidt, *Changes in Personal, Social, and Intellectual Behavior in*

Children Originally Classified as Feebleminded, Psychological Monographs (Washington, D.C.: American Psychological Association, 1946).

18. Charles M. Locurto, *Sense and Nonsense About IQ* (New York: Praeger, 1991), 3–8.

19. Confessional literature discussed in James W. Trent, Jr., *Inventing the Feeble Mind* (Berkeley: University of California Press, 1995), 230–37. For history of AHRC, see in-house history by David Goode, AHRC, New York, 1999. The Fernald League is described in *Outline History of the Walter E. Fernald School* (Waltham, Mass.: Howe Library, Walter E. Fernald School, undated).

20. Interviews with Frederick Boyce and Joseph Almeida.

21. *Industrial Training for Imbecile and Moron Boys of School Age* (Waltham, Mass.: Walter E. Fernald State School, undated). Production figures from superintendent reports. Descriptions from former students, teachers, and attendants.

22. From student records.

23. From interviews with Robert and Albert Gagne, July 8, 2002, and from official records.

24. From interviews with Doris (Gagne) Perugini, February 12, 2003.

FOUR

1. Anecdotes based on interviews with Robert Gagne, Albert Gagne, Frederick Boyce, Charles Hatch, Charles Dyer, conducted 2002–3. Brazier's comments from school record of Boyce.

2. Restraint and seclusion reported monthly. Documents held by Massachusetts State Archives. Mood of institution recalled in interviews with Frederick Boyce, Charles Hatch, and Joseph Almeida.

3. Population figures and criminal incidents reported in minutes of monthly superintendent meetings with trustees. Conditions in wards recalled in interviews.

4. Trustee monthly reports, June, July, August 1952.

5. From school file of Joseph Almeida.

6. Honeysett case reported by superintendent to trustees. Recalled in interview with Kenneth Bilodeau, June 2002.

7. Interviews with Frederick Boyce, February 2002.

8. Christmas programs and scripts held by Howe Library, Walter E. Fernald State School.

9. Christmas cake incident from interviews with Joseph Almeida, Eric Johnson, Frederick Boyce, winter 2002–3.

10. School record of Frederick Boyce.

11. Incidents of abuse recounted by multiple interviews with Joseph Almeida, Frederick Boyce, and Lawrence Gomes in 2002–3. Confirmed in school records of Joseph Almeida.

12. From interviews with Frederick Boyce, February 15, 2002, and records of Frederick Boyce. For shock therapy, see J. Campbell, "Electric Shock Treatment in Mental Deficiency," *American Journal of Mental Deficiency,* July 1953.

FIVE

1. Clemens Benda, "Biological Roots of Psychiatry," *Journal of the Philadelphia Psychiatric Hospital,* February 1957. See also "A Lonely Little Genius," *Life,* March 22, 1954.

2. "Up I.Q. by Mother's Diet," *Science Digest,* June 1955, and B. Fine, "How to Raise Your Child's IQ," *Coronet,* October 1955.

3. Cyril Burt, *The Backward Child* (London: University Press, 1950), and Cyril Burt, *The Subnormal Mind* (Oxford: Oxford University Press, 1955).

4. From school records of students named.

5. Interview with Rose Terry, July 2002. Staff problems described in *Massachusetts Needs in Mental Health and the Care of the Retarded* (Boston: Commonwealth of Massachusetts, 1958).

6. Interviews with Hatch and Williams, also from state records of Williams.

7. Interviews with Frederick Boyce and Joseph Almeida. Fires are reported in monthly superintendent statements to trustees and recorded in daily logbooks, Walter E. Fernald State School. Other details confirmed by daily logbooks, Walter E. Fernald State School.

8. Interviews with Gagne, Donovan, and other former state school residents, attendants, and staff. Autopsies are recorded in monthly superintendent reports to trustees as well as in monthly reports of clinical director.

9. Interviews with former residents and staff. See also Walter E. Fernald, *Adolescent Boys and Employees,* a letter dated February 26, 1922, held by Howe Library, Walter E. Fernald State School.

10. Interviews with Almeida and others.

11. Interview with Adrian Blake, M.D., former Fernald doctor, and interview with Patrick O'Callaghan, former Fernald staff supervisor.

12. Interview with Albert Gagne, January 2003.

13. Walter E. Fernald State School records of Frederick Boyce.

14. NARC eventually became the Association for the Help of Retarded Children (AHRC). See in-house history by David Goode, AHRC, New York, 1999.

15. TV appearance noted in trustee minutes. Work of commission is noted in various editions of *Report of Special Commission Established to Make an Investigation and Study Relative to the Training Facilities Available for Retarded Children, Commonwealth of Massachusetts, 1953–1958.*

16. Superintendent's monthly reports to trustees.

17. Conditions at various schools are described throughout Edward Shorter, *The Kennedy Family and the Story of Retardation* (Philadelphia: Temple University Press, 2000). Dybwad's experiences are told in his "Address to the New York State Commission on Quality Care for the Mentally Disabled," May 12, 1988, Albany, N.Y., held by Samuel Gridley Howe Library, Waltham, Mass.

18. From Fernald record of Frederick Boyce.

19. Interviews with persons named, including Mary Mone, Frederick Boyce, Kenneth Bilodeau, and William Duca. Interviews conducted 2002. Information also

confirmed by school records of Boyce and by daybooks of the institution, which record escapes, retrievals, and admissions to various wards.

SIX

1. Incidents of break-ins, fires, and thefts recorded in daily logbooks of the institution. Changes in commitment procedures noted in Malcolm J. Farrell, "The Present Status of Defective Delinquency in Massachusetts," *American Journal of Mental Deficiency,* January 1955.

2. Benda memo to Farrell dated November 19, 1954.

3. Interviews with Pichey, May 2002. The new federal law was the Vocational Rehabilitation Act Amendments of 1954. The arrival of new social workers and the development of a social work field clinic are noted in annual reports of the Fernald School trustees. These reports also describe the development of individual and group psychology programs at the school.

4. References to the introduction of Thorazine are found in annual reports of the school, 1955–56. Thorazine's history is recounted throughout Peter Breggin, *Toxic Psychiatry* (New York: St. Martin's Press, 1994). See also G. R. Sprogis et al., "Comparison Study on Thorazine and Serpasil in the Mental Defective," *American Journal of Mental Deficiency,* April 1957, and C. H. Martin, "The Clinical Use of Reserpine and Chlorpromazine in the Care of the Mentally Deficient," *American Journal of Mental Deficiency,* September 1957.

5. For Benda's LSD work, see Clemens Benda, "Biological Roots of Psychiatry," *Journal of the Philadelphia Psychiatric Hospital,* February 1957. Rinkel's work is recounted in Max Rinkel, *Biological Treatment of Mental Illness* (New York: L. C. Paige, 1958), and in Martin A. Lee and Bruce Shlain, *Acid Dreams* (New York: Grove Press, 1986).

6. Interviews with Frederick Boyce, Charles Hatch, and Edward Hatch, conducted in 2002.

7. Interviews and superintendent reports.

8. Escapes and retrievals recorded in Fernald logbooks. Remarks from trustees' reports.

9. D. H. Thomas, "Impressions of the Social Problem of Mental Retardation in the United States of America," *American Journal of Mental Deficiency* (September 1958): 354.

10. Staff hires and the activities and backgrounds of physicians were presented in monthly superintendent reports to trustees.

11. Incidents recalled in interviews with named sources and other witnesses.

12. From school files of Albert Gagne, Joseph Almeida, Frederick Boyce, and others.

13. Runaways noted in daily logs of attendants. Sources also include interviews with Charles Hatch and others.

14. Logbooks recorded departures and arrivals of students. For examples of local media coverage of civil rights, see "Little Rock Walkout," *Waltham News Tribune,* October 3, 1957. For example of news media coverage of Sputnik, see "Dog Alive in Space," *Boston Globe,* November 4, 1957.

15. The story of the insurrection was recounted by Charles Hatch, Joseph Almeida, Pat O'Callaghan, Kenneth Bilodeau, and others who were witnesses. Further documentation is found in state files of those who participated. Other sources include reports by Malcolm Farrell to trustees and newspaper accounts, including "Delinquents Riot Nearly Seven Hours," *Boston Herald,* November 5, 1957, and "15 Rioting Fernald Patients Defy Authorities for 7 Hours," *Waltham News Tribune,* November 5, 1957.

16. "Delinquents Riot Nearly Seven Hours."

17. Proceedings were described by Farrell in minutes of trustees meetings following riot. Further detail is contained in state files of Joseph Almeida.

18. Interviews with Charles Hatch and Joseph Almeida.

19. Special report to trustees, Massachusetts State Archives.

20. Interviews with Frederick Boyce and Joseph Almeida, July 2002.

SEVEN

1. The story of Frederick Boyce's escape and return was obtained through interviews with Boyce and those who lived in his ward at the time, including Lawrence Nutt and Joseph Almeida. Further documentation is provided in Boyce's official files, and in reports from the daily logbooks of attendants.

2. From interviews with O'Callaghan and Boyce. Record of events is also available in social work notes section of Frederick Boyce's state files.

3. From Robert and Albert Gagne's Fernald files.

4. From reports of Abigail Bacon and interviews.

5. Mildred Brazier's retirement address to the Fernald trustees, 1959.

6. Interviews with Gomes and Almeida. Reports also found in Almeida's state files.

7. Robert Wallace, "A Lifetime Thrown Away by a Mistake 50 Years Ago," and "Mental Homes Wrongly Held Thousands Like Mayo Buckner," *Life,* March 24, 1958. See also "Letters to the Editor," *Life,* March 31, 1958.

8. See in-house history by David Goode, AHRC, New York, 1999.

EIGHT

1. Fernald's move to the new Wechsler test was reported to the trustees by the psychology department. Test scores are reported in residents' state files. Further details from interviews with Gomes.

2. Jones results and reports are in the Benda Manuscript Collection, Box 5, Countway Library Archives, Harvard University.

3. Boston Radio Archives, MIT Laboratory for Computer Science, Cambridge, Mass.

4. Photos and notices of dances available in Howe Library Collection, Walter E. Fernald State School.

5. Notice of check sent to Frederick Boyce contained in his state files. Also based on interview with Frederick Boyce.

6. Based on interviews with former residents and staff.

7. Benda letter to Sargent Shriver dated November 5, 1958.

8. *Report of Special Commission Established to Make an Investigation and Study Relative to the Training Facilities Available for Retarded Children, Commonwealth of Massachusetts, 1953–1958.*

9. Details of counseling sessions are related in social service notes to Frederick Boyce's school files and the files of others. Further information from interviews. Formation of counseling program reported and updated in trustees' monthly reports.

10. Interview with Joseph Almeida and others. Crimes reported in attendants' notes to superintendent and in daily logbooks.

11. Based on interviews with Adrian Blake, M.D., Frederick Boyce, and others who were members of Boys Town.

12. Ball game was reported in Fernald newsletter and recalled by various participants.

13. Abigail Bacon reports to files of Robert and Albert Gagne. Also based on interviews with the Gagne brothers.

14. Letters and narrative of events found in Robert Williams's state school file. Also based on interview.

15. Interviews with Charles Hatch, his father, and his brother Edward.

16. Frederick Boyce's parole and settlement in Waltham documented in his state file. Events also related in interviews.

17. Parole of Fernald residents noted in superintendent's reports.

18. Editorship of this journal is reported annually in its editions.

19. Joseph Almeida's life after Fernald described by him and several contemporaries.

NINE

1. Derrick Z. Jackson, "Out of the Rubble, Resolve," *Boston Globe,* March 2, 1997. See also David Kruh, *Always Something Doing: Boston's Infamous Scollay Square* (Phoenix: Futech, 1999), and Nobuo Abiko, "Boston Shrinks, Suburbia Gains," *Christian Science Monitor,* July 2, 1960.

2. Alexander Von Hoffman, "High-Rise Hellholes," *American Prospect,* April 9, 2001; Nobuo Abiko, "Plan Bared for Fight on Delinquency," *Christian Science Monitor,* July 8, 1960; "Court Action Looms in Teen Beatings," *Christian Science Monitor,* September 9, 1960; Robert Y. Ellis, "Youth Facilities Broadened," *Christian Science Monitor,* March 18, 1961.

3. Interview with Frederick Boyce, February 2001.

4. Frederick Boyce's life in the Back Bay is reconstructed using interviews with him, Joseph Almeida, and Robert Catalano.

5. Ibid.

6. "Hooting Crowd Harasses Police at Bookie Raid," *Christian Science Monitor,* May 16, 1960; "New Police Set-up Hits Bookies Hard," *Christian Science Monitor,* September 14, 1960; "Boston Police Join T-Men in Bookie Raid," *Christian Science Monitor,* September 29, 1960.

7. Based on interviews with Frederick Boyce and Mason's wife, Dale Mason.

8. Interview with Frederick Boyce, February 2001.

9. From state file on Frederick Boyce.

10. Interview with Adrian Blake, M.D. And Susan Kelly, *The Boston Stranglers* (New York: Kensington, 1995), 395–99.

11. From Commonwealth of Massachusetts criminal record of Joseph Almeida; also from interviews with Joseph Almeida.

12. Wayne S. Sellman, *Project 100,000—Testimony and Report on the Study of Vietnam Era Low Aptitude Military Recruits,* Subcommittee on Oversight and Investigation, Committee on Veterans Affairs, U.S. Congress, Washington, D.C., February 28, 1990.

13. Interviews with Albert, Doris, and Karen Gagne.

14. Interview with Doris (Gagne) Perugini, February 2003.

15. Letter and other information contained in Richard Williams's official case files from Walter E. Fernald State School.

16. Interview with Charles Hatch.

17. Burton Blatt, *Christmas in Purgatory* (Boston: Allyn & Bacon, 1966).

18. From interview with Florence Little Kelly and also from Kelly's personal letters.

19. Rivera's work is cited in James W. Trent, Jr., *Inventing the Feeble Mind* (Berkeley: University of California Press, 1995), 258–59. The Florida case is noted in Edward Shorter, *The Kennedy Family and the Story of Mental Retardation* (Philadelphia: Temple University Press, 2000), 25.

20. Interview with Louis Frankowski. See also "Massachusetts Gaining in Its Care for Retarded," *New York Times,* January 4, 1986.

21. Institutional census figures from Massachusetts Department of Mental Retardation. For national statistics, see also *State of the States on Developmental Disabilities* (Sacramento: Association of Regional Centers, 2002).

22. Interviews with Frederick Boyce and Joseph Almeida.

TEN

1. Interviews with Frederick Boyce and Robert Catalano.

2. Ibid.

3. Interview with Joseph Almeida. Documentation included in Almeida's Fernald file. Oath of Allegiance from Almeida's personal possessions.

4. Interviews with Frederick Boyce.

5. Story of their relationship from interviews with Abra Glenn-Allen Figueroa and Frederick Boyce.

ELEVEN

1. Radio coverage of the Science Club story prompted by Scott Allen, "Radiation Used on Retarded; Postwar Experiments Done at Fernald School," *Boston Globe,* December 26, 1993.

2. Interviews with Frederick Boyce and John Dougherty; also based on video recording of the report aired on WBZ.

3. Interview with Sandra Marlow and the intermediary for her contacts with the press, attorney Daniel Burnstein.

4. Documentation of Marlow's contacts with federal authorities held in her personal collection.

5. The relationship and investigations done by Joseph Almeida and Sandra Marlow were confirmed by both in interviews in 2002.

6. "Dear Parent" letter held by Howe Library, Walter E. Fernald State School. Other documentation included in monthly reports of the experiment sent to Atomic Energy Commission and in internal memoranda of the state school. All on file at Howe Library.

7. Interviews with Sandra Marlow and Daniel Burnstein, 2002–3.

8. Allen, "Radiation Used on Retarded."

9. Ibid.

10. Scott Allen and Dolores Kong, "'50 Memo Warned Radiation Tests Would Suggest Nazism," *Boston Globe,* December 28, 1993.

11. Brian McGrory and Sean P. Murphy, "Inmates Used in '60s Drug Test; Leary Directed Study on Crime," *Boston Globe,* January 1, 1994; also Alison Bass, "Mass. Mental Patients Given LSD in '50s, Researcher Says," *Boston Globe,* January 4, 1994.

12. Ana Puga, "Energy Chief Fighting a Culture of Secrecy," *Boston Globe,* December 31, 1993. See also "Clinton Supports Release of Test Data," *Boston Globe,* January 2, 1994, and Ana Puga, "Independent Panel to Weigh Propriety of Radiation Tests; Energy Dept. Says Fernald Records Lost," *Boston Globe,* January 12, 1994.

13. John W. Mashek, "CIA Launches Radiation Probe; Records Are Checked on Cold War Era Tests on Humans," *Boston Globe,* January 5, 1994; also David Armstrong, "State Names 10 to Panel on Fernald Experiments," *Boston Globe,* January 6, 1994.

14. Interviews with Sandra Marlow, Frederick Boyce, and Joseph Almeida. Copy of Science Club document from Boyce.

15. From videotape recording of hearing. See also Dolores King, "Kennedy Opens Radiation Inquiry," *Boston Globe,* January 14, 1994.

16. Interviews with Joseph Almeida and Frederick Boyce.

17. Interviews with David White-Lief, commission member.

18. Obituary, "Clemens E. Benda, Psychiatrist, 76," *New York Times,* April 25, 1975.

19. A 1959 Benda report to administration on current research at Fernald notes Grant B-933 from the National Institute of Neurological Diseases and Blindness for bone marrow study. Also a list of Fernald residents who participated in the study was in Benda's papers.

20. Interviews with Sandra Marlow, Frederick Boyce, and Joseph Almeida.

21. Fernald log of escapes.

22. Clemens Benda, "Biological Roots of Psychiatry," *Journal of the Philadelphia Psychiatric Hospital,* February 1957.

23. Max Rinkel, *Biological Treatment of Mental Illness* (New York: L. C. Paige, 1958). See also Martin A. Lee and Bruce Shlain, *Acid Dreams* (New York: Grove Press, 1986), and Michael Hollingshead, *The Man Who Turned On the World* (London: Blond & Briggs, 1973).

24. Letter dated November 26, 1963, to Benda from E. Taylor Parks, U.S. Department of State, references Benda's request for concentration camp "autopsy material."

25. Clemens Benda, "Come, Come Now," *New York Review of Books*, November 30, 1972. Benda letter on fluoridation from his personal files. Quotes on his salesmanship from Chip Brown, "The Science Club Serves Its Country," *Esquire*, December 1994.

26. From videotape recording of *Rolanda* program.

27. Interview with David White-Lief.

28. *A Report on the Use of Radioactive Materials in Human Subject Research That Involved Residents of State-Operated Facilities Within the Commonwealth of Massachusetts from 1943 through 1973* (Boston: Commonwealth of Massachusetts, 1994).

29. Scott Allen, "Fernald Radiation Report Released, Finds No Evidence of Physical Harm," *Boston Globe*, May 10, 1994.

TWELVE

1. From videotape recording of Frederick Boyce and Wallace Cummins.

2. Michael D'Antonio, "Atomic Guinea Pigs," *New York Times Magazine*, August 31, 1997.

3. Cummins & Brown case files, interviews.

4. Report of Accidents to Patients, April 20, 1957.

5. Anecdotes from interviews with Frederick Boyce, Robert Williams, and Joseph Almeida. Information on Earl Badgett from Connecticut crime and corrections records.

6. Boyce's testimony from his written record. Work of the advisory committee is recorded in *Final Report of Advisory Committee on Human Radiation Experiments* (Washington, D.C.: U.S. Government Printing Office, October 1995).

7. Based on interviews with Michael Mattchen and Science Club members present.

8. From John Kelly's written reports to Attorney Cummins. Also from interviews with Cummins and Kelly.

9. "U.S. Apologizes to Thousands It Exposed to Radiation," *Boston Globe*, October 4, 1995.

10. From case files of Jeff Petrucelly and Mike Mattchen. See also Scott Allen, "MIT, Quaker Oats, Fernald Doctors Face $60M Lawsuit Over Tests," *Boston Globe*, December 6, 1995.

11. State files for Albert Gagne.

12. Interviews with attorneys and others at the proceedings.

13. From case files and interviews with Frederick Boyce, Joseph Almeida, Robert Gagne, Albert Gagne, and Charles Dyer, 2002–3.

AFTERWORD

1. Bill Baskerville, "Virginia Gov Apologizes for Eugenics Law," Associated Press, Richmond, Va., May 2, 2002.

2. Videotape recording of high school class meeting.

3. Barbara Meltz, "From Worry to Wisdom; Facing a Learning Disability," *Boston Globe,* October 10, 2002.

4. Lobotomies listed in superintendent's reports, 1946 and 1947. Possible names for experiment in note in Benda files dated November 5, 1950.

5. Notes on Eighteenth Experiment, dated November 1952.

6. Notes on Twelfth Experiment, dated May 1952.

7. David J. Rothman et al., *The Willowbrook Wars* (New York: HarperCollins, 1994).

8. Sacha Pfeifer, "Fernald Lauded and Lamented; It's a Last Hope or a Warehouse," *Boston Globe,* October 25, 1998. David Armstrong, "State Probe Finds Rape, Possible Abuse at Fernald School," *Boston Globe,* January 8, 1997.

9. Julie Sullivan, "Governor Candidates Ask Apology on Eugenics," *Oregonian,* July 31, 2002, and "Iowa Sued on Behalf of 5 Orphans Taught to Stutter," Associated Press, Iowa City, Iowa, April 6, 2003.

10. Emily Sweeney, "Closing the Fernald Center," *Boston Globe,* February 28, 2003.

INDEX

O'Leary, Hazel, 245, 262
Oliver, Mina Boyce, 22, 29, 32, 197–200, 223–24, 229
Oregon, 278

Paige, John C., 27
parole program, 122, 129–30, 210, 211
    Albert Gagne in, 130, 156–57, 166, 172–73
    Fred Boyce in, 156, 167–68, 175–78
    Joey Almeida and, 169–70, 179, 180–81
    residences in, 183, 184
    Robert Gagne in, 172–73
    supervision in, 176, 180
Passing of the Great Race, The (Grant), 8–9, 17
Peace Corps, 235–36
Pennachio, Daniel, 201
Pennsylvania, 55
Perugini, Doris Gagne, 37, 38, 73–76
    attempted bullying of, 74
    escape of, 74–76, 209, 210
    in foster care, 73–74
    institutionalization of, 73–74
    later life of, 209–10
    as live-in maid, 75
Petrucelly, Jeffrey, 269–71, 279
photographic exposés, 212–214
phytates, 56–57
Pichey, Raymond, 122
Piersall, Jimmy, 122
"pig piling," 92
Plath, Sylvia, 51
plutonium injections experiment, 258
pneumonia, deaths from, 54
polio vaccines, human testing of, 55
Polk State School, 55
Portland Oregonian, 278
print shop, 71
Progressivism, 8, 12
prostitution, 8, 13
    White Tower and, 187, 190–91
psychosurgery, *see* lobotomies
psychotherapy, 167–68, 178
punishments, 45–48
    beatings, 4, 54, 71, 82, 92, 100, 111, 126, 168

in behavior-modification program, 213
boxing matches, 92
bread rolling, 77, 79–80
electroshock treatments, 96
face-slapping, 92–94, 123
knee-bending, 67
kneeling on bed irons, 46–47, 67, 123
North Building placement as, 124–25, 149–55, 257
"pig piling," 92
public nakedness, 77–78
red cherries (welts), 4
spankings, 4
striking skulls with spoon, 45, 46, 92
striking with keys, 45
testicle-yanking, 45–46
toilet bowl dunking, 22, 92
urine dowsing, 4, 44
see also Ward 22

Quaker Oats Company, 56, 238, 243, 248, 253, 268–71
see also nutrition studies

Race Betterment Foundation, 8
racial differences, 7, 8–9, 15, 16–17
racial hygiene, 16
radiation research, 238–71
    of AEC, 244–45
    atomic bomb tests in, 241, 257–58
    brain tumor experiment in, 263
    Clinton's apology for, 266–68, 271
    compensation for subjects of, 245, 262, 267
    declassified documents on, 244–45, 258, 262
    environmentally released radiation in, 245, 258
    informed consent in, 245
    lawsuits resulting from, 257–58
    media coverage of, 244–45, 253, 270
    number of experiments in, 262
    plutonium injections in, 258
    presidential advisory committee on, 245, 262–64, 266–68
    radionuclide injections in, 270

secrecy of, 241, 245, 262, 264, 270
total body irradiation in, 262–63
see also nutrition studies
RADLAW, 256–57
Ready, Joseph, 175
Red Sox baseball team, 56, 122
reform movement, 110–12, 121, 161,
  166–67, 178, 212–15
retarded children, 3–4, 16, 24, 33, 43,
  66, 82, 98
  abuse and neglect of, 111–12, 113
  changing attitudes toward, 110–12,
    121, 161, 166–67
  Home Boys, 32, 61, 86, 118, 131–32,
    144, 145
Ribak, Sidney, 211
Rinkel, Max, 123, 251
Rivera, Geraldo, 213–14
Roman Empire, as Nordic, 9
Romney, Mitt, 278
Roosevelt, Theodore, 8, 9

Sabin, Albert, 55
Salk, Jonas, 55
Samans, Beverly, 201
Sasser, Alfred, 160–61
Schmidt, Bernadine, 64–65
Schwartz Key shop, 195
Science Club, 56–58, 238–71, 275
  informed consent not given by, 239,
    252, 259, 263
  members of, 56, 246–47, 248, 250,
    259–62, 264–66, 269, 270–71,
    273, 277, 279
  recommended compensation of, 255
  rewards offered to, 56, 57, 58, 238,
    254, 256, 263, 266
  see also nutrition studies
Science Club lawsuit, 256–71, 275
  formal apologies denied in, 271
  plaintiffs in, 259–62, 264–66, 269,
    270–71
  presidential advisory committee
    hearings and, 262–64
  settlement of, 257, 271, 273, 274
  statute of limitations in, 255, 264,
    270
Science Digest, 99

scientific research, 52–58
  hepatitis in, 277
  lobotomies in, 277
  polio vaccines in, 55
  stuttering experiment in, 278
  testosterone in, 277
  Thorazine in, 123
  see also Benda, Clemens E.; nutrition
    studies; radiation research
Seguin, Edouard, 36
selective breeding, 7
self-destructive behavior, 83, 101–2
Settipane, Mr., 129
sexual abuse, 51, 52, 109, 111, 278
  by female attendants, harm caused by,
    107–8
  of Joey Almeida, 93–94, 108, 158–60,
    162, 192, 227
  of Larry Nutt, 58
"sexual quarantine," 13
sexual relationships, 106–8, 158–60
  with female attendants, 261, 273–74
  homosexual, 108, 109, 158–59, 165,
    192
Shattuck, Gordon, 270–71
Shaw, Regina, 128, 154–55, 170, 171,
  175–76
Shriver, Eunice and Sargent, 167
Simpson, O. J., 268
Skeels, Harold, 15–16
Skinner, B. F., 213, 266
Sneider, Priscilla, 273–74
social sciences, 7, 8, 15–16
Sonoma State Home, 55
Soviet Union, 53
  education in, 98
  spacecraft of, 133, 134
  see also Cold War
special education programs, 110, 212,
  214–15, 276
speech habits, 41, 146–47, 192, 231
Stanford-Binet IQ Test, 10–11, 36–37
State Boys:
  academic deficiencies of, 69–70, 80,
    116, 162, 164, 248, 261
  adult attention craved by, 68–69
  adult role models sought by, 70, 166
  baseball game won by, 171